WHY BEAUTY IS TRUTH

WHY BEAUTY IS TRUTH

A History of Symmetry

IAN STEWART

BASIC
BOOKS

A Member of the Perseus Books Group
New York

Published by Basic Books
A Member of the Perseus Books Group

Designed by Jeff Williams

Library of Congress Cataloging-in-Publication Data

Stewart, Ian, 1945-
  Why beauty is truth : a history of symmetry / Ian Stewart.
    p. cm.
  ISBN-13: 978-0-465-08236-0
  ISBN-10: 0-465-08236-X
  1. Symmetry—History. I. Title.

Q172.5.S95S744 2007
539.7'25—dc22

2006038274

10 9 8 7 6 5 4 3 2 1

When old age shall this generation waste,
Thou shalt remain, in midst of other woe
Than ours, a friend to man, to whom thou say'st,
"Beauty is truth, truth beauty,"—that is all
Ye know on earth, and all ye need to know.

—JOHN KEATS, *Ode on a Grecian Urn*

CONTENTS

PREFACE

The date is 13 May 1832. In the dawn mist, two young Frenchmen face each other, pistols drawn, in a duel over a young woman. A shot is fired; one of the men falls to the ground, fatally wounded. He dies two weeks later, from peritonitis, aged 21, and is buried in the common ditch—an unmarked grave. One of the most important ideas in the history of mathematics and science very nearly dies with him.

The surviving duelist remains unknown; the one who died was Évariste Galois, a political revolutionary and a mathematical obsessive whose collected works fill a mere sixty pages. Yet Galois left a legacy that revolutionized mathematics. He invented a language to describe symmetry in mathematical structures, and to deduce its consequences.

Today that language, known as "group theory," is in use throughout pure and applied mathematics, where it governs the formation of patterns in the natural world. Symmetry also plays a central role at the frontiers of physics, in the quantum world of the very small and the relativistic world of the very large. It may even provide a route to the long-sought "Theory of Everything," a mathematical unification of those two key branches of modern physics. And it all began with a simple question in algebra, about the solutions of mathematical equations—finding an "unknown" number from a few mathematical clues.

Symmetry is not a number or a shape, but a special kind of *transformation*—a way to move an object. If the object looks the same after being transformed, then the transformation concerned is a symmetry. For instance, a square looks the same if it is rotated through a right angle.

This idea, much extended and embellished, is fundamental to today's scientific understanding of the universe and its origins. At the heart of Albert Einstein's theory of relativity lies the principle that the laws of physics should be the same in all places and at all times. That is, the laws should be symmetric with respect to motion in space and the passage of time. Quantum physics tells us that everything in the universe is built from a collection of very tiny "fundamental" particles. The behavior of these particles is governed by mathematical equations—"laws of nature"—and those laws again possess symmetry. Particles can be transformed mathematically into quite different particles, and these transformations also leave the laws of physics unchanged.

These concepts, and more recent ones at the frontiers of today's physics, could not have been discovered without a deep mathematical understanding of symmetry. This understanding came from pure mathematics; its role in physics emerged later. Extraordinarily useful ideas can arise from purely abstract considerations—something that the physicist Eugene Wigner referred to as "the unreasonable effectiveness of mathematics in the natural sciences." With mathematics, we sometimes seem to get more out than we put in.

Starting with the scribes of ancient Babylon and ending with the physicists of the twenty-first century, *Why Beauty Is Truth* tells how mathematicians stumbled upon the concept of symmetry, and how an apparently useless search for what turned out to be an impossible formula opened a new window on the universe and revolutionized science and mathematics. More broadly, the story of symmetry illustrates how the cultural influence and historical continuity of big ideas can be brought into sharp relief by occasional upheavals, both political and scientific.

The first half of the book may seem at first sight to have nothing to do with symmetry and precious little to do with the natural world. The reason is that symmetry did not become a dominant idea by the route one might expect, through geometry. Instead, the profoundly beautiful and indispensable concept of symmetry that mathematicians and physicists use today arrived via algebra. Much of this book, therefore, describes the search for solutions of algebraic equations. This may sound technical, but the quest is a gripping one, and many of the key players led unusual and dramatic lives. Mathematicians are human, even though they are often lost in abstract thought. Some of them may let logic rule their lives too much, but we shall see time and again that our heroes could in fact be all too human. We will see how they lived and died, read of their love affairs and duels, vicious priority disputes, sex scandals, drunkenness, and disease, and along the way we will see how their mathematical ideas unfolded and changed our world.

Beginning in the tenth century BCE and reaching its climax with Galois in the early nineteenth century, the narrative retraces the step-by-step conquest of equations—a process that eventually ground to a halt when mathematicians tried to conquer the so-called "quintic" equation, involving the fifth power of the unknown. Did the methods break down because there was something fundamentally different about the quintic equation? Or might there be similar, yet more powerful methods that would lead to formulas for its solution? Were mathematicians stuck because of a genuine obstacle, or were they just being obtuse?

It is important to understand that solutions to quintic equations were known to exist. The question was, can they always be represented by an algebraic formula? In 1821 the young Norwegian Niels Henrik Abel proved that the quintic equation cannot be solved by algebraic means. His proof, however, was rather mysterious and indirect. It proved that no general solution is possible, but it did not really explain *why*.

It was Galois who discovered that the impossibility of solving the quintic stems from the symmetries of the equation. If those symmetries pass the Galois test, so to speak—meaning that they fit together in a very specific way, which I will not explain just yet—then the equation can be solved by an algebraic formula. If the symmetries do not pass the Galois test, then no such formula exists.

The general quintic equation cannot be solved by a formula *because it has the wrong kind of symmetries.*

This epic discovery created the second theme of this book: that of a *group*—a mathematical "calculus of symmetry." Galois took an ancient mathematical tradition, algebra, and reinvented it as a tool for the study of symmetry.

At this stage of the book, words like "group" are unexplained jargon. When the meaning of such words becomes important to the story, I will explain them. But sometimes we just need a convenient term to keep track of various items of baggage. If you run into something that looks like jargon but is not immediately discussed, then it will be playing the role of a useful label, and the actual meaning won't matter very much.

Sometimes the meaning will emerge anyway if you keep reading. "Group" is a case in point, but we won't find out what it means until the middle of the book.

Our story also touches upon the curious significance of particular numbers in mathematics. I am not referring to the fundamental constants of physics but to mathematical constants like π (the Greek letter pi). The speed of light, for instance, might in principle be anything, but it happens to be 186,000 miles per second in our universe. On the other hand, π is slightly larger than 3.14159, and nothing in the world can change that value.

The unsolvability of the quintic equation tells us that like π, the number 5 is also very unusual. It is the smallest number for which the associated symmetry group fails the Galois test. Another curious example concerns the sequence of numbers 1, 2, 4, 8. Mathematicians discovered a series of extensions of the ordinary "real" number concept to complex numbers and then to things called quaternions and octonions. These are constructed from two copies of the real numbers, four copies, and eight copies, respectively. What comes next? A natural guess is 16, but in fact there are *no* further sensible extensions of the number system. This fact is remarkable and deep. It tells us that there is something special about the number 8, not in any superficial sense, but in terms of the underlying structure of mathematics itself.

In addition to 5 and 8, this book features appearances by several other numbers, most notably 14, 52, 78, 133, and 248. These curious numbers are the dimensions of the five "exceptional Lie groups," and their influence pervades the whole of mathematics and much mathematical physics. They are key characters in the mathematical drama, while other numbers, seemingly little different, are mere bit players.

Mathematicians discovered just how special these numbers are when modern abstract algebra came into being at the end of the nineteenth century. What counts is not the numbers themselves but the role they play in the foundations of algebra. Associated with each of these numbers is a mathematical object called a Lie group with unique and remarkable properties. These groups play a fundamental role in modern physics, and they appear to be related to the deep structure of space, time, and matter.

That leads to our final theme: fundamental physics. Physicists have long wondered why space has three dimensions and time one—why we live in a four-dimensional space-time. The theory of superstrings, the most recent attempt to unify the whole of physics into a single coherent set of laws, has led physicists to wonder whether space-time might have extra "hidden" dimensions. This may sound like a ridiculous idea, but it has good historical precedents. The presence of additional dimensions is probably the least objectionable feature of superstring theory.

A far more controversial feature is the belief that formulating a new theory of space and time depends mainly on the *mathematics* of relativity and quantum theory, the two pillars on which modern physics rests. Unifying these mutually contradictory theories is thought to be a mathematical exercise rather than a process requiring new and revolutionary experiments. Mathematical beauty is expected to be a prerequisite for physical truth. This could be a dangerous assumption. It is important not to lose sight of the physical world, and whatever theory finally emerges from today's deliberations cannot be exempt from comparison with experiments and observations, however strong its mathematical pedigree.

At the moment, though, there are good reasons for taking the mathematical approach. One is that until a really convincing combined theory is formulated, no one knows what experiments to perform. Another is that mathematical symmetry plays a fundamental role in both relativity and quantum theory, two subjects where common ground is in short supply, so we should value whatever bits of it we can find. The possible structures of space, time, and matter are determined by their symmetries, and some of the most important possibilities seem to be associated with exceptional structures in algebra. Space-time may have the properties it has because mathematics permits only a short list of special forms. If so, it makes sense to look at the mathematics.

Why does the universe seem to be so mathematical? Various answers have been proposed, but I find none of them very convincing. The symmetrical relation between mathematical ideas and the physical world, like the symmetry between our sense of beauty and the most profoundly important mathematical forms, is a deep and possibly unsolvable mystery. None of us can say *why* beauty is truth, and truth beauty. We can only contemplate the infinite complexity of the relationship.

1

THE SCRIBES OF BABYLON

Across the region that today we call Iraq run two of the most famous rivers in the world, and the remarkable civilizations that arose there owed their existence to those rivers. Rising in the mountains of eastern Turkey, the rivers traverse hundreds of miles of fertile plains, and merge into a single waterway whose mouth opens into the Persian Gulf. To the southwest they are bounded by the dry desert lands of the Arabian plateau; to the northeast by the inhospitable ranges of the Anti-Taurus and Zagros Mountains. The rivers are the Tigris and the Euphrates, and four thousand years ago they followed much the same routes as they do today, through what were then the ancient lands of Assyria, Akkad, and Sumer.

To archaeologists, the region between the Tigris and Euphrates is known as Mesopotamia, Greek for "between the rivers." This region is often referred to, with justice, as the cradle of civilization. The rivers brought water to the plains, and the water made the plains fertile. Abundant plant life attracted herds of sheep and deer, which in turn attracted predators, among them human hunters. The plains of Mesopotamia were a Garden of Eden for hunter-gatherers, a magnet for nomadic tribes.

In fact, they were so fertile that the hunter-gatherer lifestyle eventually became obsolete, giving way to a far more effective strategy for obtaining food. Around 9000 BCE, the neighboring hills of the Fertile Crescent, a little to the north, bore witness to the birth of a revolutionary technology: agriculture. Two fundamental changes in human society followed hard on its heels: the need to remain in one location in order to tend the crops, and the possibility of supporting large populations. This combination led to the invention of the city, and in Mesopotamia we can still find archaeological remains of some of the earliest of the world's great city-states: Nineveh, Nimrud, Nippur, Uruk, Lagash, Eridu, Ur, and above all, Babylon, land of the Hanging Gardens and the Tower of Babel. Here, four millennia ago, the agricultural revolution led inevitably to an organized society, with all the associated trappings of government, bureaucracy, and military power. Between 2000 and 500 BCE the civilization that is loosely termed "Babylonian" flourished on the banks of the Euphrates. It is named for its capital city, but in the broad sense the term "Babylonian" includes Sumerian and Akkadian cultures. In fact, the first known mention of Babylon occurs on a clay tablet of Şargon of Akkad, dating from around 2250 BCE, although the origin of the Babylonian people probably goes back another two or three thousand years.

We know very little about the origins of "civilization"—a word that literally refers to the organization of people into settled societies. Nevertheless, it seems that we owe many aspects of our present world to the ancient Babylonians. In particular, they were expert astronomers, and the twelve constellations of the zodiac and the 360 degrees in a circle can be traced back to them, along with our sixty-second minute and our sixty-minute hour. The Babylonians needed such units of measurement to practice astronomy, and accordingly had to become experts in the time-honored handmaiden of astronomy: mathematics.

Like us, they learned their mathematics at school.

"What's the lesson today?" Nabu asked, setting his packed lunch down beside his seat. His mother always made sure he had plenty of bread and meat—usually goat. Sometimes she put a piece of cheese in for variety.

"Math," his friend Gamesh replied gloomily. "Why couldn't it be law? I can *do* law."

Nabu, who was good at mathematics, could never quite grasp why his fellow students all found it so difficult. "Don't you find it boring, Gamesh, copying all those stock legal phrases and learning them by heart?"

Gamesh, whose strengths were stubborn persistence and a good memory, laughed. "No, it's easy. You don't have to *think*."

"That's precisely why I find it boring," his friend said, "whereas math—"

"—is impossible," Humbaba joined in, having just arrived at the Tablet House, late as usual. "I mean, Nabu, what am I supposed to do with *this?*" He gestured at a homework problem on his tablet. "I multiply a number by itself and add twice the number. The result is 24. What is the number?"

"Four," said Nabu.

"Really?" asked Gamesh. Humbaba said, "Yes, I know, but how do you *get* that?"

Painstakingly, Nabu led his two friends through the procedure that their math teacher had shown them the week before. "Add half of 2 to 24, getting 25. Take the square root, which is 5—"

Gamesh threw up his hands, baffled. "I've never really grasped that stuff about square roots, Nabu."

"Aha!" said Nabu. "Now we're getting somewhere!" His two friends looked at him as if he'd gone mad. "Your problem isn't solving equations, Gamesh. It's square roots!"

"It's both," muttered Gamesh.

"But square roots come first. You have to master the subject one step at a time, like the Father of the Tablet House keeps telling us."

"He also keeps telling us to stop getting dirt on our clothes," protested Humbaba, "but we don't take any notice of—"

"That's different. It's—"

"It's *no good!*" wailed Gamesh. "I'll never become a scribe, and my father will wallop me until I can't sit down, and mother will give me that pleading look of hers and tell me I've got to work harder and think of the family. But I can't get math into my head! Law, I can remember. It's fun! I mean how about 'If a gentleman's wife has her husband killed on account of another man, they shall impale her on a stake'? That's what I call worth learning. Not dumb stuff like square roots." He paused for breath and his hands shook with emotion. "Equations, numbers—why do we *bother?*"

"Because they're useful," replied Humbaba. "Remember all that legal stuff about cutting off slave's ears?"

"Yeah!" said Gamesh. "Penalties for assault."

"Destroy a common man's eye," prompted Humbaba, "and you must pay him—"

"One silver *mina,*" said Gamesh.

"And if you break a slave's bone?"

"You pay his master half the slave's price in compensation."

Humbaba sprung his trap. "So, if the slave costs sixty shekels, then you have to be able to work out half of sixty. If you want to practice law, you need math!"

"The answer's thirty," said Gamesh immediately.

"See!" yelled Nabu. "You *can* do math!"

"I don't *need* math for that, it's obvious." The would-be lawyer flailed the air, seeking a way to express the depth of his feelings. "If it's about the real world, Nabu, yes, I can do the math. But not artificial problems about square roots."

"You need square roots for land measurement," said Humbaba.

"Yes, but I'm not studying to become a tax collector, my father wants me to be a scribe," Gamesh pointed out. "Like him. So I don't see why I have to learn all this math."

"Because it's useful," Humbaba repeated.

"I don't think that's the real reason," Nabu said quietly. "I think it's all about truth and beauty, about getting an answer and knowing that it's right." But the looks on his friends' faces told him that they weren't convinced.

"For me it's about getting an answer and knowing that it's wrong," sighed Gamesh.

"Math is important because it's true and beautiful," Nabu persisted. "Square roots are fundamental for solving equations. They may not be much use, but that doesn't matter. They're important for themselves."

Gamesh was about to say something highly improper when he noticed the teacher walking into the classroom, so he covered his embarrassment with a sudden attack of coughing.

"Good morning, boys," said the teacher brightly.

"Good morning, master."

"Let me see your homework."

Gamesh sighed. Humbaba looked worried. Nabu kept his face expressionless. It was better that way.

Perhaps the most astonishing thing about the conversation upon which we have just eavesdropped—leaving aside that it is complete fiction—is that it took place around 1100 BCE, in the fabled city of Babylon.

Might have taken place, I mean. There is no evidence of three boys named Nabu, Gamesh, and Humbaba, let alone a record of their conversation. But human nature has been the same for millennia, and the factual background to my tale of three schoolboys is based on rock-hard evidence.

We know a surprising amount about Babylonian culture because their records were written on wet clay in a curious wedge-shaped script called cuneiform. When the clay baked hard in the Babylonian sunshine, these inscriptions became virtually indestructible. And if the building where the clay tablets were stored happened to catch fire, as sometimes happened—well, the heat turned the clay into pottery, which would last even longer.

A final covering of desert sand would preserve the records indefinitely. Which is how Babylon became the place where written history begins. The story of humanity's understanding of symmetry—and its embodiment in a systematic and quantitative theory, a "calculus" of symmetry every bit as powerful as the calculus of Isaac Newton and Gottfried Wilhelm Leibniz—begins here too. No doubt it might be traced back further, if we had a time machine or even just some older clay tablets. But as far as recorded history can tell us, it was Babylonian mathematics that set humanity on the path to symmetry, with profound implications for how we view the physical world.

Mathematics rests on numbers but is not limited to them. The Babylonians possessed an effective notation that, unlike our "decimal" system (based on powers of ten), was "sexagesimal" (based on powers of sixty). They knew about right-angled triangles and had something akin to what we now call the Pythagorean theorem—though unlike their Greek successors, the mathematicians of Babylon seem not to have supported their empirical findings with logical proofs. They used mathematics for the higher purpose of astronomy, presumably for agricultural and religious reasons, and also for the prosaic tasks of commerce and taxation. This dual role of mathematical thought—revealing order in the natural world and assisting in human affairs—runs like a single golden thread throughout the history of mathematics.

What is most important about the Babylonian mathematicians is that they began to understand how to solve equations.

Equations are the mathematician's way of working out the value of some unknown quantity from circumstantial evidence. "Here are some known facts about an unknown number: deduce the number." An equation, then, is a kind of puzzle, centered upon a number. We are not told what this number is, but we are told something useful about it. Our task is to solve the puzzle by finding the unknown number. This game may seem somewhat divorced from the geometrical concept of symmetry, but in mathematics, ideas discovered in one context habitually turn out to illuminate very different contexts. It is this interconnectedness that gives mathematics such intellectual power. And it is why a number system invented for commercial reasons could also inform the ancients about the movements of the planets and even of the so-called fixed stars.

The puzzle may be easy. "Two times a number is sixty: what is the number we seek?" You do not have to be a genius to deduce that the unknown number is thirty. Or it may be much harder: "I multiply a number by itself and add 25: the result is ten times the number. What is the number we seek?" Trial and error may lead you to the answer 5—but trial and error is an inefficient way to answer puzzles, to solve equations. What if we change 25 to 23, for example? Or 26? The Babylonian mathematicians disdained trial and error, for they knew a much deeper, more powerful secret. They knew a rule, a standard procedure, to solve such equations. As far as we know, they were the first people to realize that such techniques existed.

The mystique of Babylon stems in part from numerous Biblical references. We all know the story of Daniel in the lion's den, which is set in Babylon during the reign of King Nebuchadnezzar. But in later times, Babylon became almost mythical, a city long vanished, destroyed beyond redemption, that perhaps had never existed. Or so it seemed until roughly two hundred years ago.

For thousands of years, strange mounds had dotted the plains of what we now call Iraq. Knights returning from the Crusades brought back souvenirs dragged from the rubble—decorated bricks, fragments of undecipherable inscriptions. The mounds were clearly the ruins of ancient cities, but beyond that, little was known.

In 1811, Claudius Rich made the first scientific study of the rubble mounds of Iraq. Sixty miles south of Baghdad, beside the Euphrates, he surveyed the entire site of what he soon determined must be the remains of Babylon, and hired workmen to excavate the ruins. The finds included bricks, cuneiform tablets, beautiful cylinder seals that produced raised words and pictures when rolled over wet clay, and works of art so majestic that whoever carved them must be ranked alongside Leonardo da Vinci and Michelangelo.

Even more interesting, however, were the smashed cuneiform tablets that littered the sites. We are fortunate that those early archaeologists recognized their potential value, and kept them safe. Once the writing had been deciphered, the tablets became a treasure-trove of information about the lives and concerns of the Babylonians.

The tablets and other remains tell us that the history of ancient Mesopotamia was lengthy and complex, involving many different cultures and states. It is customary to employ the word "Babylonian" to refer to them all, as well as to the specific culture that was centered upon the city of Babylon. However, the heart of Mesopotamian culture moved repeatedly, with Babylon both coming into, and falling out of, favor. Archaeologists divide Babylonian history into two main periods. The Old Babylonian period runs from about 2000 to 1600 BCE, and the Neo-Babylonian period runs from 625 to 539 BCE. In between are the Old Assyrian, Kassite, Middle Assyrian, and Neo-Assyrian periods, when Babylon was ruled by outsiders. Moreover, Babylonian mathematics continued in Syria, throughout the period known as Seleucid, for another five hundred years or more.

The culture itself was much more stable than the societies in which it resided, and it remained mostly unchanged for some 1200 years, sometimes temporarily disrupted by periods of political upheaval. So any particular aspect of Babylonian culture, other than some specific historical event, probably came into existence well before the earliest known record. In particular, there is evidence that certain mathematical techniques, whose first surviving records date to around 600 BCE, actually existed far earlier. For this reason, the central character in this chapter—an imaginary scribe to whom I shall give the name Nabu-Shamash and whom we have already met during his early training in the brief vignette about three school friends—is deemed to have lived sometime around 1100 BCE, being born during the reign of King Nebuchadnezzar I.

All the other characters that we will meet as our tale progresses were genuine historical figures, and their individual stories are well documented. But among the million or so clay tablets that have survived from ancient Babylon, there is little documented evidence about specific individuals other than royalty and military leaders. So Nabu-Shamash has to be a pastiche based on plausible inferences from what we have learned about everyday Babylonian life. No new inventions will be attributed to him, but he will encounter all those aspects of Babylonian knowledge that play a role in the story of symmetry. There is good evidence that all Babylonian scribes underwent a thorough education, with mathematics as a significant component.

Our imaginary scribe's name is a combination of two genuine Babylonian names, the scribal god Nabu and the Sun god Shamash. In Babylonian culture it was not unusual to name ordinary people after gods, though perhaps two god-names would have been considered a bit extreme. But for narrative reasons we are obliged to call him something more specific, and more atmospheric, than merely "the scribe."

When Nabu-Shamash was born, the king of Babylonia was Nebuchadnezzar I, the most important monarch of the Second Dynasty of Isin. This was *not* the famous Biblical king of the same name, who is usually referred to as Nebuchadnezzar II; the Biblical king was the son of Nabopolassar, and he reigned from 605 to 562 BCE.

Nebuchadnezzar II's reign represented the greatest flowering of Babylon, both materially and in regional power. The city also flourished under his earlier namesake, as Babylon's power extended to encompass Akkad and the mountainous lands to the north. But Akkad effectively seceded from Babylon's control during the reigns of Ahur-resh-ishi and his son Tiglath-Pileser I, and it strengthened its own security by taking action against the mountain and desert tribes that surrounded it on three sides. So Nabu-Shamash's life began during a stable period of Babylonian history, but by the time he became a young man, Babylon's star was beginning to wane, and life was becoming more turbulent.

Nabu-Shamash was born into a typical "upper-class" household in the Old City of Babylon, not far from the Libil-hegalla canal and close to the justly famed Ishtar Gate, a ceremonial entrance decorated with colored ceramic bricks in fanciful forms—bulls, lions, even dragons. The road through the Ishtar Gate was impressive, reaching a width of 20 meters; it was paved with limestone flags on top of a bed of asphalt, with a brick foundation. Its name was "May the enemy not have victory"—rather typical of Babylon's main street names—but it is generally known as the Processional Way, being used by the priests to parade the god Marduk through the city when ceremony so decreed.

The family home was built of mud brick, with walls six feet thick to keep out the sun. The external walls had few openings—mainly a doorway at street level—and rose to a height of three stories, with lighter materials, mainly wood, being used for the top floor. The family owned many slaves, who performed routine household tasks. Their quarters, along with the kitchen, were to the right of the entrance. The family rooms were to the left: a long living room, bedrooms, and a bathroom. There was no bathtub in Nabu-Shamash's time, though some have survived from other eras. Instead, a slave would pour water over the bather's head and body, approximating a modern shower. A central courtyard opened to the sky, and toward the back were storerooms.

Nabu-Shamash's father was an official in the court of a king, name unknown, whose reign preceded Nebuchadnezzar I. His duties were largely bureaucratic: he was responsible for administering an entire district, ensuring that law and order were maintained, that the fields were properly irrigated, and that all necessary taxes were collected and paid. Nabu-Shamash's father had also been trained as a scribe, because literacy and numeracy were basic skills for anyone in the Babylonian equivalent of the civil service.

According to a decree attributed to the god Enlil, every man should follow in his father's footsteps, and Nabu-Shamash was expected to do just that. However, scribal abilities also opened up other career paths, notably that of priest, so his training paved the way to a choice of professions.

We know what Nabu-Shamash's education was like because extensive records, written in Sumerian by people who were trained as scribes, have survived from roughly the period concerned. These records make it plain that Nabu-Shamash was fortunate in his choice of parentage, for only the sons of the well-to-do could hope to enter the scribal schools. In fact, the quality of Babylonian education was so high that foreign nobles sent their sons to the city to be educated.

The school was called the Tablet House, presumably referring to the clay tablets used for writing and arithmetic. It had a head teacher, referred to as the "Expert" and as the "Father of the Tablet House." There was a class teacher, whose main task was to make the boys behave themselves; there were specialist teachers in Sumerian and mathematics. There were prefects, called "Big Brothers," whose job included keeping order. Like all students, Nabu-Shamash lived at home and went to school during the day, for around 24 days each 30-day month. He had three days off for recreation, and a further three for religious festivals.

Nabu-Shamash began his studies by mastering the Sumerian language, especially its written form. There were dictionaries and grammatical texts to be studied, and long lists to be copied—legal phrases, technical terms, names. Later, he progressed to mathematics, and it was then that his studies became central to our tale.

What did Nabu-Shamash learn? For everyone but philosophers, logicians, and professional mathematicians who are being pedantic, a number is a string of digits. Thus the year in which I write this sentence is 2006, a string of four digits. But as the pedants will jump to remind us, this string of digits is not the number at all but only its notation, and a rather sophisticated form of notation at that. Our familiar decimal system employs just ten digits, the symbols 0 through 9, to represent any number, however large. An extension of that system also permits the representation of very small numbers; more to the point, it permits the representation of numerical measurements to very high levels of precision. Thus the speed of light, according to the best current observations, is approximately 186,282.397 miles per second.

We are so familiar with this notation that we forget how clever it is—and how difficult to grasp when we first encounter it. The key feature on which all else rests is this: the numerical value of a symbol, such as 2, depends on where it is placed relative to the other symbols. *The symbol 2 does not have a fixed meaning independent of its context.* In the number representing the speed of light, the digit "2" immediately before the decimal point does indeed mean "two." But the other occurrence of "2" in that number means "two hundred." In the date 2006, that same digit means "two thousand."

We would be exceedingly unhappy to have a system of writing in which the meaning of a letter depended on where it occurred in a word. Imagine, for instance, what reading would be like if the two a's in "alphabet" had totally different meanings. But positional notation for numbers is so convenient and powerful that we find it hard to imagine that anyone really used any other method.

It was not always thus. Our present notation dates back no more than 1500 years, and was first introduced into Europe a little more than 800

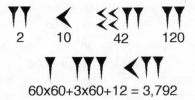

Babylonian base–60 numerals.

years ago. Even today, different cultures use different symbols for the same decimal digits—look at any Egyptian banknote. But ancient cultures wrote numbers in all sorts of strange ways. The most familiar to us is probably the Roman system, in which 2006 becomes MMVI. In ancient Greek it would be $\overline{\beta}\zeta$. In place of our 2, 20, 200, and 2000, the Romans wrote II, XX, CC, and MM, and the Greeks wrote β, κ, σ, and $\overline{\beta}$.

The Babylonians were the earliest known culture to use something akin to our positional notation. But there was one significant difference. In the decimal system, every time a digit is moved one place to the left, its numerical value is multiplied by ten. So 20 is ten times 2, and 200 is ten times 20. In the Babylonian system, each move to the left multiplies a number by sixty. So "20" would mean 2 times 60 (120 in our notation) and "200" would mean 2 times 60 times 60 (7200 in our notation). Of course, they didn't use the same "2" symbol; they wrote the number "two" using two copies of a tall, thin wedge symbol, as shown in the figure above. Numbers from one to nine were written by grouping that many copies of the tall wedge. For numbers greater than nine, they added another symbol, a sideways wedge, which denoted the number ten, and they used groups of these symbols to denote twenty, thirty, forty, and fifty. So, for instance, our "42" was four sideways wedges followed by two tall wedges.

For reasons we can only guess at, this system stopped at 59. The Babylonians did not group six sideways wedges to make 60. Instead, they reverted to the tall thin wedge previously used to mean "one," and used it to mean "one times sixty." Two such wedges meant 120. But they might also mean "two." Which meaning was intended had to be inferred from context, and from the position of the symbols relative to each other. For example, if there were two tall wedges, a space, and two more tall wedges, then the first group meant "one hundred twenty" and the second "two"— much as the two symbols 2 in our 22 mean twenty and two respectively.

This method extended to much larger numbers. A tall wedge could mean 1, or 60, or $60 \times 60 = 3,600$, or $60 \times 60 \times 60 = 216,000$, and so on. The three bottom groups in the figure indicate $60 \times 60 + 3 \times 60 + 12$, which we would write as 3,792. A big problem here is that the notation has some ambiguities. If all you see is two tall wedges, does this mean 2, 60×2, or $60 \times 60 \times 2$? Does a sideways wedge followed by two tall ones mean $12 \times 60 + 2$ or $12 \times 60 \times 60 + 2$, or even $10 \times 60 \times 60 + 2 \times 60$? By Alexander the Great's time, the Babylonians had removed these ambiguities by using a pair of diagonal wedges to indicate that no number occurred in a given slot; in effect, they had invented a symbol for zero.

Why did the Babylonians use this sexagesimal system rather than the familiar decimal system? They may have been influenced by a useful feature of the number 60: its large variety of divisors. It is divisible exactly by the numbers 2, 3, 4, 5, and 6. It is also divisible by 10, 12, 15, 20, and 30. This feature is rather pleasant when it comes to sharing things, such as grain or land, among several people.

A final feature may well have been decisive: the Babylonian method of measuring time. It seems that they found it convenient to divide a year into 360 days, although they were excellent astronomers and knew that 365 was closer, and 365¼ closer still. The lure of the arithmetical relationship $360 = 6 \times 60$ was too strong. Indeed, when referring to time, the Babylonians suspended the rule that moving symbols one slot to the left multiplied their value by sixty, and replaced that by six, so that what should have meant 3,600 was actually interpreted as 360.

This emphasis on 60 and 360 still lingers today, in our use of 360 degrees in a full circle—one degree per Babylonian day—and in the 60 seconds in a minute and 60 minutes in an hour. Old cultural conventions have incredible staying power. I find it amusing that in this age of spectacular computer graphics, moviemakers still date their creations in Roman numerals.

Nabu-Shamash would have learned all of this, except the "zero" sign, at an early stage of his education. He would have become adept at impressing thousands of tiny cuneiform wedges into damp clay at speed. And just as today's students grapple with the transition from whole numbers to fractions and decimals, Nabu-Shamash would eventually have been faced with the Babylonian method for representing numbers like one-half, or one-third, or the more complicated subdivisions of unity dictated by the brutal realities of astronomical observations.

To avoid spending whole afternoons drawing wedges, scholars represent cuneiform numbers with a mixture of old and new. They write the decimal numbers depicted in the successive groups of wedges, using commas to separate them. So the final group in the figure would be written 1,3,12. This convention saves a lot of expensive typesetting and is easier to read, so we'll go along with the scholars.

How would a Babylonian scribe have written the number "one-half"?

In our own arithmetic, we solve this problem two different ways. We either write the number as a fraction, ½, or introduce the famous "decimal point" and write it as 0.5. The fractional notation is more intuitive and came earlier historically; decimal notation is more difficult to grasp, but it lends itself better to computation because the symbolism is a natural extension of the "place-value" rules for whole numbers. The symbol 5 in 0.5 means "5 *divided by* 10," and in 0.05 it means "5 divided by 100." Moving a symbol one place to the left multiplies it by 10; moving it one place to the right divides it by 10. All very sensible and systematic.

As a result, decimal arithmetic is just like whole-number arithmetic, except that you have to keep track of where the decimal point goes.

The Babylonians had the same idea, but in base 60. The fraction ½ should be some number of copies of the fraction 1/60. Clearly the right number is 30/60, so they wrote "one-half" as 0;30, where scholars use the semicolon to denote the "sexagesimal point," which in cuneiform notation was again a matter of spacing. The Babylonians managed some fairly advanced calculations: for example, their value for the square root of 2 was 1;24,51,10, which differs from the true value by less than one part in a hundred thousand. They used this precision to good advantage in both theoretical mathematics and astronomy.

The most exciting technique that Nabu-Shamash would have been taught, as far as our central theme of symmetry is concerned, is the solution of quadratic equations. We know quite a lot about Babylonian methods for solving equations. Of the roughly one million Babylonian clay tablets known to exist, about five hundred deal with mathematics. In 1930, the orientalist Otto Neugebauer recognized that one of these tablets demonstrated a complete understanding of what today we call quadratic equations. These are equations that involve an unknown quantity and its square, together with various specific numbers. Without the square, the equation would be called "linear," and such equations are the easiest to solve. An equation that involves the cube of the unknown (multiply it by itself, then multiply that by the unknown again) is called "cubic." The Babylonians seem to have possessed a clever method for finding approximate solutions to certain types of cubic equation, based on numerical tables. All that we are certain of, however, are the tables themselves. We can only infer what they were used for, and cubic equations are most likely. But the tablets Neugebauer studied make it plain that the Babylonian scribes had mastered the quadratic.

A typical one, which dates back about 4000 years, asks, "Find the side of a square if the area minus the side is 14,30." This problem involves the square of the unknown (the area of the square) as well as the unknown itself. In other words, it asks the reader to solve a quadratic equation. The same tablet rather offhandedly provides the answer: "Take half of 1, which is 0;30. Multiply 0;30 by 0;30, which is 0;15. Add this to 14,30 to get 14,30;15. This is the square of 29;30. Now add 0;30 to 29;30. The result is 30, the side of the square."

What's going on here? Let's write the steps in modern notation.

- Take half of 1, which is 0;30.   ½

- Multiply 0;30 by 0;30, which is 0;15.   ¼

- Add this to 14,30 to get 14,30;15.   870¼

- This is the square of 29;30.   870¼ = (29½) × (29½)

- Now add 0;30 to 29;30.   29½ + ½

- The result is 30, the side of the square.   30

The most complicated step is the fourth, which finds a number (it is 29½) whose square is 870¼. The number 29½ is the *square root* of 870¼. Square roots are the main tool for solving quadratics, and when mathematicians tried to use similar methods to solve more complicated equations, modern algebra was born.

Later we will interpret this problem using modern algebraic notation. But it is important to realize that the Babylonians did not employ an algebraic formula as such. Instead, they described a specific *procedure,* in the form of a typical example, that led to an answer. But they clearly knew that *exactly the same procedure would work if the numbers were changed.*

In short, they knew how to solve quadratics, and their method—though not the form in which they expressed it—was the one we use today.

How did the Babylonians discover their method for solving quadratics? There is no direct evidence, but it seems likely that they came across it by thinking geometrically. Let's take an easier problem that leads to the same recipe. Suppose we find a tablet that says, "Find the side of a square if the area plus two of the sides is 24." In more modern terms, the square of the unknown plus twice the unknown equals 24. We can represent this question as a picture:

Geometric picture of a quadratic equation.

Here the vertical dimension of the square and rectangle to the left of the equal sign corresponds to the unknown, and the small squares are of unit size. If we split the tall rectangle in half and glue the two pieces onto the square, we get a shape like a square with one corner missing. The picture suggests that we should "complete the square" by adding in the missing corner (shaded square) to both sides of the equation:

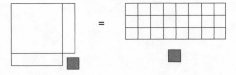

Completing the square.

Now we have a square on the left and 25 unit squares on the right. Rearrange those into a 5 × 5 square:

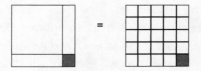

Now the solution is obvious.

Thus the unknown plus one, squared, equals five squared. Taking square roots, the unknown plus one equals five—and you don't have to be a genius to deduce that the unknown is four.

This geometric description corresponds precisely to the Babylonian method for solving quadratics. The more complicated example from the tablet uses exactly the same recipe. The tablet only states the recipe and doesn't say where it comes from, but the geometric picture fits other circumstantial evidence.

2

THE HOUSEHOLD NAME

Many of the greatest mathematicians of the ancient world lived in the Egyptian city of Alexandria, a city whose origins lie among five substantial oases to the west of the Nile, out in the Western Desert. One of them is Siwa, notable for its salt lakes, which grow during the winter and shrink in the summer heat. The salt contaminates the soil and creates major headaches for archaeologists because it is sucked up into the ancient stone and mud-brick remains and slowly destroys the fabric of the buildings.

The most popular tourist site in Siwa is Aghurmi, a former temple dedicated to the god Amun. So holy was Amun that his main aspect was entirely abstract, but he became associated with a more physical entity, the provenance of the god Re, the Sun. Constructed during the 26th Dynasty, the temple of Amun at Siwa was the home of a famous oracle that is particularly associated with two major historical events.

The first is the destruction of the army of Cambyses II, a Persian king who conquered Egypt. It is said that in 523 BCE, planning to use the oracle of Amun to legitimate his rule, Cambyses sent a military force into the Western Desert. The army reached Bahariya Oasis but was destroyed in a sandstorm on its way to Siwa. Many Egyptologists suspect that the "lost army of Cambyses" may be mythical, but in 2000 a team from Helwan University, looking for oil, found bits of cloth, metal, and human remains in the area, and suggested that these might be the remains of the lost army.

The second event, two centuries later, is historical fact: a fateful visit to Siwa by Alexander the Great, who was after exactly the same thing as Cambyses.

Alexander was the son of King Philip II of Macedon. Philip's daughter Cleopatra of Macedon married King Alexander of Epirus, and Philip was assassinated during the proceedings. The killer may have been Philip's homosexual lover Pausanias, who was upset because the king had not done anything about some complaint or other that Pausanias had made. Or the murder may have been a Persian plot set up by Darius III. If so, it backfired, because the Macedonian army immediately proclaimed Alexander king, and the 20-year-old monarch famously went on to conquer most of the known world. Along the way, in 332 BCE, he conquered Egypt without a fight.

Intent on cementing this conquest with an endorsement of his credentials as pharaoh, Alexander made a pilgrimage to Siwa to ask the oracle whether he was a god. He visited the oracle alone, and on his return announced its verdict: yes, the oracle had confirmed that he truly was a god. This verdict became the primary source of his authority. Later, rumors claimed that the oracle had revealed him to be the son of Zeus.

It is not clear whether the Egyptians were convinced by this rather flimsy evidence or whether, given Alexander's control of a substantial army, they found it prudent to go along with his story. Perhaps they were fed up with the rule of the Persians and considered Alexander the lesser of two evils—he had been welcomed with open arms by the former Egyptian capital of Memphis for precisely that reason. Whatever the truth behind the history, from that time on, the Egyptians venerated Alexander as their king.

On the way to Siwa, fascinated by an area of the country lying between the Mediterranean Sea and the lake that came to be known as Mareotis, Alexander decided to have a city constructed there. The city, which he modestly named Alexandria, was designed by Donocrates, a Greek architect, after a basic plan sketched by Alexander himself. The city's birth has been dated by some to 7 April 331 BCE; this date is disputed by others, but it must be close to 334 BCE. Alexander never saw his creation; his next visit to the area was to be buried there.

So, at least, goes the time-honored legend, but the truth is probably more complex. It now appears that much of what later became Alexandria already existed when Alexander arrived. Egyptologists discovered long ago that many inscriptions are not all that trustworthy. The great Temple at Karnak, for instance, is riddled with cartouches of Ramses II. But much of it was actually constructed by his father, Seti I, and traces—

not always faint—of the father's inscriptions can be seen beneath those carved for Ramses. Such usurpation was commonplace, and was not even considered disrespectful. In contrast, "defacing" a predecessor's reliefs—hacking out the pharaoh's face—was most definitely disrespectful, intentionally depriving that predecessor of his place in the afterlife by destroying his very identity.

Alexander had his name carved all over the buildings of ancient Alexandria. He had his name carved, so to speak, on the city itself. Where other pharaohs usurped the odd building or monument, Alexander usurped an entire city.

Alexandria became a major seaport, connected by branches of the Nile and a canal to the Red Sea and thence to the Indian Ocean and the Far East. It became a center of learning, with a celebrated library. And it was the birthplace of one of the most influential mathematicians in history: the geometer Euclid.

We know much more about Alexander than we do about Euclid—even though Euclid's long-term influence on human civilization was arguably greater. If there can be such a thing as a household name in mathematics, "Euclid" is it. Although we know little about Euclid's life, we know a lot about his works. For several centuries, mathematics and Euclid were pretty much synonymous throughout the Western world.

Why did Euclid become so well known? There have been greater mathematicians, and more significant ones. But for close to two thousand years Euclid's name was known to every student of mathematics across the whole of Western Europe, and to a lesser extent in the Arab world as well. He was the author of one of the most famous mathematics texts ever written: the *Elements of Geometry* (usually shortened to *Elements*). When printing was invented, this work was among the first books to appear in printed form. It has been published in over a thousand different editions, a number exceeded only by the Bible.

We know slightly more about Euclid than we do about Homer. He was born in Alexandria around 325 BCE and died in about 265 BCE.

Having said that, I am uncomfortably aware that I already need to backtrack. That Euclid existed and was sole author of the *Elements* is only one of three theories. The second is that he existed but did not write the *Elements,* at least not on his own. He may have been the leader of a team of mathematicians who collectively produced the *Elements*. The third theory—far more contentious but within the bounds of possibility—is that the team existed, but much like the group of mostly French, mostly young mathematicians who wrote under the name "Nicolas Bourbaki" in the mid-twentieth century, they took "Euclid" as a collective pseudonym. Nevertheless, the most likely story seems to be that Euclid existed, that he was one person, and that he composed the *Elements* himself.

This does not mean that Euclid discovered all of the mathematics contained within his book's pages. What he did was to collect and codify a substantial part of ancient Greek mathematical knowledge. He borrowed from his predecessors and he left a rich legacy for his successors, but he also stamped his own authority on the subject. The *Elements* is generally described as a geometry book, but it also deals with number theory and a kind of prototypical algebra—all of it presented in geometrical guise.

Of Euclid's life we know very little. Later commentators included a few snippets of information in their works, none of which modern scholars can substantiate. They tell us that Euclid taught in Alexandria, and it is usual to infer that he was born in that city, but we don't actually know that. In 450 AD, in an extensive commentary on Euclid's mathematics written more than seven centuries after his death, the philosopher Proclus wrote:

> Euclid . . . put together the *Elements,* arranging in order many of Eudoxus's theorems, perfecting many of Theaetetus's, and also bringing to irrefutable demonstration the things which had been only loosely proved by his predecessors. This man lived in the time of the first Ptolemy; for Archimedes, who followed closely upon the first Ptolemy, makes mention of Euclid, and further they say that Ptolemy once asked him if there were a shorter way to study geometry than the *Elements,* to which he replied that there was no royal road to geometry. He is therefore younger than Plato's circle, but older than Eratosthenes and Archimedes; for these were contemporaries, as Eratosthenes somewhere says. In his aim he was a Platonist, being in sympathy with this philosophy, whence he made the end of the whole *Elements* the construction of the so-called Platonic figures.

The treatment of some topics in the *Elements* provides indirect but compelling evidence that Euclid must at some point have been a student at Plato's Academy in Athens. Only there, for example, could he have learned about the geometry of Eudoxus and Theaetetus. As for his character, all we have are some fragments from Pappus, who described him as "most fair and well disposed towards all who were able in any measure to advance mathematics, careful in no way to give offence, and although an exact scholar, not vaunting himself." A few anecdotes survive, such as one told by Stobaeus. One of Euclid's students asked him what he would get through an understanding of geometry. Euclid called his slave and said, "Give him a coin, since he must make a profit from what he learns."

The Greek attitude to mathematics was very different from that of the Babylonians or the Egyptians. Those cultures saw mathematics largely in practical terms—although "practical" could mean aligning shafts through a pyramid so that the *ka* of the dead pharaoh could be launched in the direction of Sirius. For some Greek mathematicians, numbers were not tools occasionally employed in support of mystical beliefs, but the very core of those beliefs.

Aristotle and Plato tell of a cult, centered on Pythagoras, that flourished around 550 BCE and that viewed mathematics, especially number, as the basis of the whole of creation. They developed mystical ideas about the harmony of the universe, based in part on the discovery that harmonious notes on a stringed instrument are related to simple mathematical patterns. If a string produces a certain note, a string of half the length produces a note one octave higher—the most harmonious of all intervals. They investigated various number patterns, in particular polygonal numbers, formed by arranging objects in polygonal patterns. For instance, the "triangular numbers" 1, 3, 6, and 10 are formed from triangles, and the "square numbers" 1, 4, 9, and 16 are formed from squares:

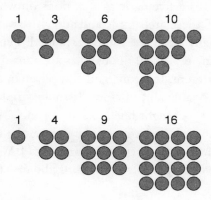

Triangular and square numbers.

Pythagoreanism embraced some nutty numerology—it considered 2 to be male and 3 female, for example—but the view that the deep structure of nature is mathematical survives today as the basis of most theoretical science. Although later Greek geometry was less mystical, the Greeks generally saw mathematics as an end in itself, more a branch of philosophy than a tool.

There are reasons to believe that this does not tell the whole story. It is well established that Archimedes, who may have been a pupil of Euclid's, employed his mathematical abilities to design powerful machines and engines of war. There survive a tiny number of intricate Greek mechanisms whose cunning design and precision manufacture hint at a well-developed tradition of craftsmanship—an ancient version of "applied mathematics." Perhaps the best-known example is a machine found in the sea near the small island of Antikythera that appears to be a calculating device for astronomical phenomena built from a complex tangle of interlocking cogwheels.

Euclid's *Elements* certainly fits this rarefied view of Greek mathematics—possibly because that view is largely based on the *Elements*. The book's main emphasis is on logic and proof, and there is no hint of practical applications. The most important feature of the *Elements,* for our story, is not what it contains but what it does not.

Euclid made two great innovations. The first is the concept of proof. Euclid refuses to accept any mathematical statement as being true unless it is supported by a sequence of logical steps that deduces it from statements already known to be true. The second innovation recognizes that the proof process must start somewhere, and that these initial statements cannot be proved. So Euclid states up front five basic assumptions on which all his later deductions rest. Four of these are simple and straightforward: two points may be joined by a line; any finite line can be extended; a circle can be drawn with any center and any radius; all right angles are equal.

But the fifth postulate is very different. It is long and complicated, and what it asserts is not nearly so reasonable and obvious. Its main implication is the existence of parallel lines—straight lines that never meet, but run forever in the same direction, always the same distance apart, like two sidewalks on either side of an infinitely long, perfectly straight road. What Euclid actually states is that whenever two lines cross a third, the first two

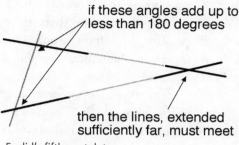

if these angles add up to
less than 180 degrees

then the lines, extended
sufficiently far, must meet

Euclid's fifth postulate.

lines must meet on whichever side creates two angles that add up to less than two right angles. It turns out that this assumption is logically equivalent to the existence of exactly one line parallel to a given line and passing through a given point (not on the given line).

For centuries the fifth postulate was viewed as a blemish—something to be removed by deducing it from the other four, or to be replaced by something simpler and just as obvious as the others. By the nineteenth century, mathematicians understood that Euclid was absolutely right to include his fifth postulate, because they could prove that it can't be deduced from his other assumptions.

To Euclid, logical proofs were an essential feature of geometry, and proof remains fundamental to the mathematical enterprise. A statement that lacks a proof is viewed with suspicion, however much circumstantial evidence seems to favor it and however important its implications may be. Physicists, engineers, and astronomers tend to view proofs with disdain, as a kind of pedantic appendage, because they have an effective substitute: observation.

For instance, imagine an astronomer trying to calculate the movements of the Moon. He will write down mathematical equations that determine the Moon's motion, and promptly get stuck because there seems to be no way to solve the equations exactly. So the astronomer may tinker with the equations, introducing various simplifying approximations. A mathematician will worry that these approximations might have a serious effect on the answer, and will want to prove that they do not cause trouble. The astronomer has a different way to check that what he has done makes sense.

He can see whether the motion of the Moon fits his calculations. If it does, that simultaneously justifies the method (because it got the right answer) and verifies the theory (for the same reason). The logic here is not circular because if the method is mathematically invalid, then it will almost certainly fail to predict the Moon's motion.

Without the luxury of observations or experiments, mathematicians have to verify their work by its internal logic. The more important the implications of some statement are, the more important it is to make sure that the statement is true. So proof becomes even more crucial when the statement is something that everyone wants to be true, or that would have enormous implications if it *were* true.

Proofs cannot rest on thin air, and they cannot trace logical antecedents back forever. They have to start somewhere, and where they start will by definition be things that have not been—and will not be—proved. Today we call these unproved starting assumptions *axioms*. The axioms for a piece of mathematics are the rules of the game.

Anyone who objects to the axioms can change them if they wish: however, the result will be a different game. Mathematics does not assert that some statement is *true*: it asserts that if we make various assumptions, then the statement concerned must be a logical consequence. This does not imply that the axioms are unchallengeable. Mathematicians may debate whether a given axiomatic system is better than another for some purpose, or whether the system has any intrinsic merit or interest. But these discussions are not about the internal logic of any particular axiomatic game. They are about which games are worthwhile, interesting, or fun.

The consequences of Euclid's axioms—his long, carefully selected chains of logical deductions—are extraordinarily far-reaching. For example, he proves—with logic that in his day was considered impeccable—that once you agree to his axioms you necessarily must conclude that:

- The square on the hypotenuse of a right triangle is equal to the sum of the squares on the other two sides.

- There exist infinitely many prime numbers.

- There exist irrational numbers—not expressible as an exact fraction. An example is the square root of two.

- There are precisely five regular solids: the tetrahedron, cube, octahedron, dodecahedron, and icosahedron.

- Any angle can be divided exactly into two equal parts using only straightedge and compass.

- Regular polygons with 3, 4, 5, 6, 8, 10, and 12 sides can be constructed exactly using only straightedge and compass.

I have expressed these "theorems," as we call any mathematical statement that has a proof, in modern terms. Euclid's point of view was rather different: he did not work directly with numbers. Everything that we would interpret as properties of numbers is stated in terms of lengths, areas, and volumes.

The content of the *Elements* divides into two main categories. There are theorems, which tell you that something is true. And there are constructions, which tell you how to *do* something.

A typical and justly famous theorem is Proposition 47 of Book I of the *Elements,* usually known as the Pythagorean theorem. This tells us that the longest side of a right triangle bears a particular relationship to the other two sides. But it does not, without further effort or interpretation, provide a method for achieving any goal.

A construction that will be important in our story is Proposition 9 of Book I, where Euclid solves the "bisection problem" for angles. Euclid's

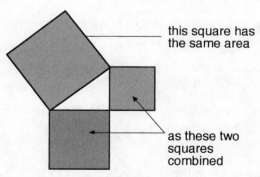

this square has
the same area

as these two
squares
combined

The Pythagorean theorem.

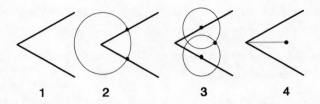

How to bisect an angle with straightedge and compass.

method of bisecting an angle is simple but clever, given the limited techniques available at this early stage of the development.

Given (1) an angle between two line segments, (2) place your compass tip where the segments meet, and draw a circle, which crosses the segments at two points, one on each (dark blobs). Now (3) draw two circles of equal radius, one centered at each of the new points. They meet in two points (only one is marked), and (4) the required bisector (dotted) runs through both of these.

By repeating this construction, you can divide an angle into four equal pieces, or eight, or sixteen—the number doubles at each step, so we obtain the powers of 2, which are 2, 4, 8, 16, 32, 64, and so on.

As I mentioned, the main aspect of *The Elements* that affects our story is not what it contains but what it doesn't. Euclid did not provide any method for:

- Dividing an angle into *three* exactly equal parts ("trisecting the angle").

- Constructing a regular 7-sided polygon.

- Constructing a line whose length is equal to the circumference of a given circle ("rectifying the circle").

- Constructing a square whose area is equal to that of a given circle ("squaring the circle").

- Constructing a cube whose volume is exactly twice that of a given cube ("duplicating the cube").

It is sometimes said that the Greeks themselves saw these omissions as flaws in Euclid's monumental work and devoted a great deal of effort to repairing them. Historians of mathematics have found very little evidence to back up these claims. In fact, the Greeks could solve all of the above problems, but they had to use methods that were not available within the Euclidean framework. All of Euclid's constructions were done with an unmarked straightedge and compass. Greek geometers could trisect angles using special curves called conic sections; they could square the circle using another special curve called a quadratrix. On the other hand, they do not seem to have realized that if you can trisect angles, you can construct a regular 7-gon. (I *do mean* 7-gon. There is an easy construction for a 9-gon, but there is also a very clever one for a 7-gon.) In fact, they apparently did not follow up the consequences of trisection at all. Their hearts seem not to have been in it.

Later mathematicians viewed Euclid's omissions in a rather different light. Instead of seeking new tools to solve these problems, they began to wonder what could be achieved with the limited tools Euclid used: straightedge and compass. (And no cheating with marks on the ruler: the Greeks knew that "neusis constructions" with sliding rulers and alignment of marks could trisect the angle effectively and accurately. One such method was devised by Archimedes.) Finding out what could or could not be done, and proving it, took a long time. By the late 1800s we finally knew that none of the above problems can be solved using straightedge and compass alone.

This was a remarkable development. Instead of proving that a particular method solved a particular problem, mathematicians were learning to prove the opposite, in a very strong form: *no* method of this-and-that kind is capable of solving such-and-such problem. Mathematicians began

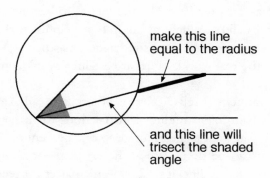

How Archimedes trisected an angle.

to learn the inherent limitations of their subject. With the fascinating twist that even as they were stating these limitations, they could prove that they genuinely *were* limitations.

In the hope of avoiding misconceptions, I want to point out some important aspects of the trisection question.

What is required is an *exact* construction. This is a very strict condition within the idealized Greek formulation of geometry, where lines are infinitely thin and points have zero size. It requires cutting the angle into three *exactly* equal parts. Not just the same to ten decimal places, or a hundred or a billion—the construction must be *infinitely* precise. In the same spirit, however, we are allowed to place the compass point with infinite precision on any point that is given to us or is later constructed; we can set the radius of the compass, with infinite precision, to equal the distance between any two such points; and we can draw a straight line that passes *exactly* through any two such points.

This is not what happens in messy reality. So is Euclid's geometry useless in the real world? No. For instance, if you do what Euclid prescribes in Proposition 9, with a real compass on real paper, you get a pretty good bisector. In the days before computer graphics, this is how draftsmen bisected angles in technical drawings. Idealization is not a flaw: it is the main reason mathematics works at all. Within the idealized model, it is possible to reason logically, because we know exactly what properties our objects have. The messy real world isn't like that.

But idealizations also have limitations that sometimes make the model inappropriate. Infinitely thin lines do not, for example, work well as painted lane markers on roads. The model has to be tailored to an appropriate context. Euclid's model was tailored to help us work out the logical dependencies among geometrical statements. As a bonus, it may also help us understand the real world, but that certainly wasn't central to Euclid's thinking.

The next comment is related, but it points in a rather different direction. There is no problem finding constructions for trisecting angles approximately. If you want to be accurate to one percent or one thousandth of a percent, it can be done. When the error is one thousandth of the thickness of your pencil line, it really doesn't matter for technical drawing.

The mathematical problem is about *ideal* trisections. Can an arbitrary angle be trisected *exactly?* And the answer is "no."

It is often said that "you can't prove a negative." Mathematicians know this to be rubbish. Moreover, negatives have their own fascination, especially when new methods are needed to prove them. Those methods are often more powerful, and more interesting, than a positive solution. When someone invents a powerful new method to characterize those things that can be constructed using straightedge and compass, and distinguish them from those that cannot, then you have an *entirely new way of thinking.* And with that come new thoughts, new problems, new solutions—and new mathematical theories and tools.

No one can use a tool that hasn't been built. You can't call a friend on your cell phone if cell phones don't exist. You can't eat a spinach soufflé if no one has invented agriculture or discovered fire. So tool-building can be at least as important as problem-solving.

The ability to divide angles into equal parts is closely related to something much prettier: constructing regular polygons.

A polygon (Greek for "many angles") is a closed shape formed from straight lines. Triangles, squares, rectangles, diamonds like this ◊, all are polygons. A circle is not a polygon, because its "side" is a curve, not a series of straight lines. A polygon is regular if all of its sides have the same length and if each pair of consecutive sides meet at the same angle. Here are regular polygons with 3, 4, 5, 6, 7, and 8 sides:

Regular polygons.

Their technical names are equilateral triangle, square, (regular) pentagon, hexagon, heptagon, and octagon. Less elegantly, they are referred to as the regular 3-gon, 4-gon, 5-gon, 6-gon, 7-gon, and 8-gon. This terminology may seem ugly, but when it becomes necessary to refer to the regular 17-sided polygon—as it shortly will—then the term "17-gon" is far more practical than "heptadecagon" or "heptakaidecagon." As for the 65,537-gon (yes, that too!)—well, you get the picture.

Euclid and his predecessors must have thought a great deal about which regular polygons can be constructed, because he offers constructions for many of them. This turned out to be a fascinating, and decidedly tricky, question. The Greeks knew how to construct regular polygons when the number of sides is

$$3, 4, 5, 6, 8, 10, 12, 15, 16, 20.$$

We now know that they *cannot* be constructed when the number of sides is

$$7, 9, 11, 13, 14, 18, 19.$$

which leaves one number in this range, 17, as yet unaccounted for. The story of the 17-gon will be told in its rightful place; it is important for more reasons than purely mathematical ones.

In discussing geometry, there is no substitute for drawing on a sheet of paper with a real straightedge and real compass. It gives you a feel for how the subject fits together. I'm going to take you through my favorite construction, for the regular hexagon. I learned it from a book my uncle gave me in the late 1950s, called *Man Must Measure,* and it's lovely:

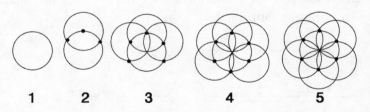

1   2   3   4   5

How to construct a regular hexagon.

Fix the radius of the compass throughout, so all circles will be of the same size. (1) Draw a circle. (2) Choose a point on it and draw a circle centered at that point. This crosses the original circle in two new points. (3) Draw circles with these points as centers, to get two more crossings. (4) Draw circles with *these* points as centers; both pass through the same new crossing point. The six points can now be connected to form a regular hexagon. It is aesthetically pleasing (though mathematically unnecessary) to complete the picture with (5): Draw a circle centered on the sixth point. Then six circles meet at the center of the original one, forming a flower shape.

Euclid used a very similar method, which is simpler but not quite so pretty, and he *proved* that it works: you can find it in Proposition 15 of Book IV.

3

THE PERSIAN POET

ake! For the Sun, who scattered into flight
The Stars before him from the Field of Night,
Drives Night along with them from Heav'n and strikes
The Sultán's Turret with a Shaft of Light.

To most of us, the name of Omar Khayyám is indelibly associated with his long ironic poem, the *Rubaiyat,* and specifically with the elegant translation into English by Edward Fitzgerald. To historians of mathematics, however, Khayyám has a greater claim to fame. He was prominent among the Persian and Arab mathematicians who took up the torch that the Greeks had dropped, and continued the development of new mathematics after scholars in Western Europe descended into the dark ages and its scholars abandoned theorem-proving for theological disputation.

Among Khayyám's great achievements is the solution, by respectable methods of Greek geometry, of cubic equations. His techniques necessarily went beyond the straightedge and compass that tacitly limit Euclidean geometry, because these tools are simply not up to the job—a fact that the Greeks strongly suspected, but could not prove because they lacked the necessary point of view, which was not geometry but algebra. But Khayyám's methods did not go *much* beyond straightedge and compass. He relied on special curves known as "conic sections" because they can be constructed by slicing a cone with a plane.

The conventional wisdom in popular science writing is that every equation halves a book's sales. If true, this is very bad news, because nobody would be able to understand some of the key themes of this book without being shown a few equations. The next chapter, for instance, is about Renaissance mathematicians' discoveries of formulas that solve any cubic or quartic equation. I can get away without showing you what the quartic formula looks like, but we really will need to take a quick look at the formula for the cubic. Otherwise, all I can tell you is something like "multiply some numbers by some other numbers and add some numbers to that, and then take the square root, then add another number and take the cube root of the result; *then* do the same thing again with slightly different numbers; finally, add the two results together. Oh, and I forgot to mention—you have to do some dividing as well."

Some writers have challenged the conventional wisdom and even written books *about* equations. They seem to be following the old showbiz saying, "If you've got a wooden leg, wave it." Now, there is a sense in which this book is about equations; but just as you can write a book about mountains without requiring your readers to climb one, you can write a book about equations without requiring your readers to solve one. Still, readers of a book about mountains probably won't understand it if they have never seen a mountain, so it really will help us both a lot if I show you a few carefully selected equations.

The ground rules, slanted heavily in your favor, are these: The word is "show." I want to you to *see* the equation. You needn't *do* anything with it. When necessary, I will pick the equation to pieces and explain which features matter for our story. I will *never* ask you to solve an equation or calculate with one. And I will do my utmost to avoid them whenever I can.

When you get to know them, equations are actually rather friendly. They are clear, concise, sometimes even beautiful. The secret truth about equations is that they are a simple, clear language for describing certain "recipes" for calculating things. When I can tell you the recipe in words, or just give you enough feel for how it goes that the details don't matter, I will. On rare occasions, though, it becomes so cumbersome to use words that I'll have to use symbols.

There are three kinds of important symbols for this book, and I'll mention two of them now. One is our old friend x, "the unknown." This symbol stands for a number that we do not yet know, but whose value we are desperately trying to find out.

The second type of symbol is little raised numbers, like ^{2} or ^{3} or ^{4}. They are instructions to multiply some other number by itself the appropriate number of times. So 5^3 means $5 \times 5 \times 5$, which is 125, and x^2 means $x \times x$, where x is our symbol for an unknown number. They are read as "squared," "cubed," "raised to the fourth power," etc., and collectively they are referred to as *powers* of the number concerned.

I haven't the foggiest idea why. They have to be called something.

Either the Babylonian method for solving quadratic equations was passed on to the ancient Greeks, or they reinvented it. Heron, who lived in Alexandria somewhere between 100 BCE and 100, discussed a typical Babylonian-style problem in Greek terminology. Around the year 100, Nichomachus, probably an Arabian hailing from Judea, wrote a book called *Introductio Arithmetica* in which he abandoned the Greek tradition of representing numbers by geometrical quantities such as lengths or areas. To Nichomachus, numbers were quantities in their own right, not lengths of lines. Nichomachus was a Pythagorean, and his work shows it: he deals only in whole numbers and their ratios, and he uses no symbols. His book became the standard arithmetic text for the next millennium.

Symbolism entered into algebra in the work of a Greek mathematician named Diophantus, sometime around 500. The only thing that we know about Diophantus is his age at death, and that comes to us via a route of dubious authenticity. A Greek collection of algebra problems contains one that reads like this: "Diophantus spent one-sixth of his life as a boy. His beard grew after a further one-twelfth. He married after another one-seventh, and his son was born five years later. The son lived to half his father's age and the father died four years after the son. How old was Diophantus when he died?"

Using that ancient algebraist's own methods, or more modern ones, you can deduce that he must have been 84. It was a good age, assuming the algebra problem is based on fact, which is questionable.

That's all we know of his life. But we know quite a bit about his books, through later copies and references in other documents. He wrote one book on polygonal numbers, and part of it survives. It is arranged in Euclidean style, proves theorems using logical arguments, and has little mathematical significance. Far more significant were the 13 books of the *Arithmetica*. Six of them are still in existence, thanks to a thirteenth-century Greek copy of an earlier copy. Four more may have surfaced in a manuscript found in Iran, but not all scholars are convinced that it traces back to Diophantus.

The *Arithmetica* is presented as a series of problems. In the preface, Diophantus says he wrote it as a book of exercises for one of his students. He used a special symbol for the unknown, and different symbols for its square and cube that seem to be abbreviations of the words *dynamis* (power) and *kybos* (cube). The notation is not very structured. Diophantus adds symbols by putting them next to each other (as we now do for multiplication) but has a special symbol for subtraction. He even has a symbol for equality, though this may have been introduced by later copyists.

Mostly, the *Arithmetica* is about solving equations. The first surviving book discusses linear equations; the other five treat various kinds of quadratic equations, often in several unknowns, and a few special cubic equations. A key feature is that the answers are always integers or rational numbers. Today we call an equation "Diophantine" if its solutions are restricted to integers or rational numbers. A typical example from the *Arithmetica* is, "Find three numbers such that their sum, and the sum of any two, is a perfect square." Try it—it's by no means easy. Diophantus's answer is 41, 80, and 320. The sum of all three is $441 = 21^2$. The sums of pairs are $41 + 80 = 121 = 11^2$, $41 + 320 = 361 = 19^2$, and $80 + 320 = 400 = 20^2$. Clever stuff.

Diophantine equations are central to modern number theory. A famous example is Fermat's "last theorem," which states that two perfect cubes, or higher powers, cannot add to form a similar power. With squares, this kind of thing is easy, and goes back to Pythagoras: $3^2 + 4^2 = 5^2$ or $5^2 + 12^2 = 13^2$. But you can't do the same with cubes, fourth powers, fifth powers, or anything higher than the square. Pierre de Fermat scribbled this conjecture (without a proof; it wasn't a theorem despite its name) in the margin of his personal copy of the *Arithmetica* around 1650. It took nearly 350 years before Andrew Wiles, a British-born number theorist living in America, proved that Fermat was right.

The historical tradition in mathematics is sometimes very long.

Algebra really arrived on the mathematical scene in 830, when the main action moved from the Greek world to the Arabic one. In that year the astronomer Mohamed ibn Musa al-Khwārizmī wrote a book called *al-Jabr w'al Muqâbala,* which translates roughly as "restoration and simplification." The words refer to standard techniques for manipulating equations so as to put them into a better form for solution. From *al-jabr* comes our word "algebra." The first Latin translation in the twelfth century bears the title *Ludus Algebrae et Almucgrabalaeque.*

Al-Khwārizmī's book contains hints of earlier influences, Babylonian and Greek, and also rests on ideas introduced in India by Brahmagupta around 600. It explains how to solve linear and quadratic equations. Al-Khwārizmī's immediate successors worked out how to solve a few special kinds of cubic. Among them are Tâbit ibn Qorra, a doctor, astronomer, and philosopher who lived in Baghdad and was a pagan, and an Egyptian named al-Hasan ibn al-Haitham, generally referred to in later Western writings as Alhazen. But the most famous of them all is Omar Khayyám.

Omar bore the full name Ghiyath al-Din Abu'l-Fath Umar ibn Ibrahim Al-Nisaburi al-Khayyámi. The word "al-Khayyámi" translates literally as "tent-maker," which some scholars believe may have been the trade of his father Ibrahim. Omar was born in Persia in 1047, and spent most of his productive life at Naishapur. You can find it in an atlas as Neyshabur, a city near Masshad in the Khorosan province of northeastern Iran, close to the border with Turkmenistan.

Legend has it that in his youth, Omar left home to study Islam and the Quran under the celebrated cleric Imam Mowaffak, who lived in Naishapur. There he struck up a friendship with two fellow students, Hasan Sabah and Nizam al-Mulk, and the three of them made a pact. If any of them became rich and famous—not unlikely for students of Mowaffak—that person would share his wealth and power with the other two.

The students completed their studies and the years flew by; the pact remained in force. Nizam traveled to Kabul. Omar, politically less ambitious, spent some time as a tent-maker—another possible explanation of the name "Al-Khayyámi." Science and mathematics became his passions, and he spent most of his spare time on them. Eventually Nizam returned, secured a position in the government, and became administrator of affairs to the sultan Alp Arslan, with an office in Naishapur.

Since Nizam was now rich and famous, Omar and Hasan claimed their rights under the pact. Nizam asked the sultan for permission to assist his friends, and when it was granted he honored the agreement. Hasan received a well-paid government job, but Omar merely wished to continue his scientific studies in Naishapur, where he would pray for the health and well-being of Nizam. His old school friend arranged for Omar to be given a government salary, to free his time for study, and the deal was done.

Hasan later tried to overthrow a senior official and lost his sinecure, but Omar continued serenely on and was appointed to a commission whose mandate was to reform the calendar. The Persian calendar was based on the movements of the sun, and the date of the first day of the new year was subject to change, which was confusing. It was just the job for a competent mathematician, and Omar applied his knowledge of mathematics and astronomy to calculate when New Year's Day should fall in any given year.

Around this time, he also penned the *Rubaiyat,* which loosely translates as "quatrains," a poetic form. A *rubai* was a four-line verse with a rather specific rhyming pattern—more accurately, one of two possible patterns—and a *rubaiyat* was a collection of verses in this form. One verse makes a clear reference to his work on reforming the calendar:

> *Ah, but my computations, People say,*
> *Reduced the Year to better Reckoning? Nay,*
> *'Twas only striking from the Calendar*
> *Unborn To-morrow and dead Yesterday.*

Omar's verses were distinctly nonreligious. Many of them praise wine and its effects. For instance:

> *And lately, by the Tavern Door agape,*
> *Came shining through the Dusk an Angel Shape*
> *Bearing a Vessel on his Shoulder; and*
> *He bid me taste of it; and 'twas—the Grape!*

There are wry allegorical references to wine, as well:

> *Whether at Naishápúr or Babylon,*
> *Whether the Cup with sweet or bitter run,*
> *The Wine of Life keeps oozing drop by drop,*
> *The Leaves of Life keep falling one by one.*

Other verses poke fun at religious beliefs. One wonders what the Sultan thought of the man he had put on retainer, and what the imam thought of the outcome of his teaching.

Meanwhile, the disgraced Hasan, having been forced to leave Naisha-pur, fell in with a gang of bandits and made use of his superior education to become its leader. In the year 1090 those bandits, under Hasan's com-mand, captured Alamut castle in the Elburz Mountains, just south of the Caspian Sea. They terrorized the region, and Hasan became notorious as the Old Man of the Mountains. His followers, known as the *Hashishiyun* for their use of the drug hashish (a very potent form of cannabis), built six mountain fortifications, from which they would emerge to kill carefully selected religious and political figures. Their name was the origin of the word "assassin." So Hasan managed to become rich and famous in his own right, as befitted a student of Mowaffak, though he was not, by this time, disposed to share his fortune with his old schoolmates.

While Omar calculated astronomical tables and worked out how to solve the cubic, Nizam pursued his political career until, in a touch of ex-quisite irony, Hasan's bandits assassinated him. Omar lived on to the age of 76, dying—so it is said—in 1123. Hasan died the following year, aged 84. The assassins continued to wreak political havoc until they were wiped out by the Mongols, who conquered Alamut in 1256.

To return to Omar's mathematics: Around 350 BCE the Greek mathe-matician Menaechmus discovered the special curves known as "conic sec-tions," which he used, scholars believe, to solve the problem of doubling the cube. Archimedes developed the theory of these curves, and Apollo-nius of Perga systematized and extended the subject in his book *Conic Sections*. What particularly interested Omar Khayyám was the Greek dis-covery that conic sections could be used to solve certain cubic equations.

Conic sections are so named because they can be obtained by slicing a cone with a plane. More properly, a *double* cone, like two ice-cream cones joined at their sharp ends. A single cone is formed by a collection of straight-line segments, all meeting at one point and passing through a suit-able circle, the "base" of the cone. But in Greek geometry you can always extend straight-line segments as far as you wish, and the result is to create a double cone.

The three main types of conic section are the ellipse, parabola, and hy-perbola. An ellipse is a closed oval curve that arises when the cutting plane passes through only one half of the double cone. (A circle is a special kind of ellipse, created when the plane is exactly perpendicular to the cone's

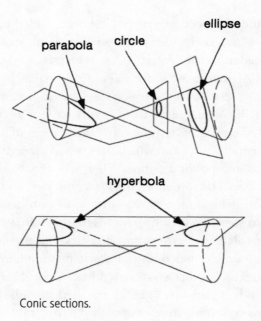

Conic sections.

axis.) A hyperbola consists of two symmetrically related open curves, which in principle extend to infinity, that arise when the cutting plane passes through both halves of the double cone. The parabola is a transitional form, a single open curve, and in this case the cutting plane must be parallel to one of the lines lying on the surface of the cone.

At great distances from the tip of the cone, the curves of a hyperbola become ever closer to two straight lines, which are parallel to the lines where a parallel plane through the tip would cut the cone. These lines are called *asymptotes*.

The Greek geometers' extensive studies of conic sections constituted their most significant area of progress beyond the ideas codified by Euclid. These curves remain vitally important in today's mathematics, but for quite different reasons from those that interested the Greeks. From the algebraic point of view, they are the next simplest curves after the straight line. They are also important in applied science. The orbits of planets in the solar system are ellipses, as Kepler deduced from Tycho Brahe's observations of Mars. This elliptical orbit is one of the observations that led Newton to formulate his famous "inverse square law" of gravity. This in turn led to the realization that some aspects of the universe exhibit clear mathematical patterns. It opened up the whole of astronomy by making planetary phenomena computable.

The majority of Omar's extant mathematics is devoted to the theory of equations. He considered two kinds of solution. The first, following the lead of Diophantus, he called an "algebraic" solution in whole numbers; a better adjective would be "arithmetic." The second kind of solution he called "geometric," by which he meant that the solution could be constructed in terms of specific lengths, areas, or volumes by geometrical means.

Making liberal use of conic sections, Omar developed geometric solutions for all cubic equations, and explained them in his *Algebra,* which he completed in 1079. Because negative numbers were not recognized in those days, equations were always arranged so that all terms were positive. This convention led to a huge number of case distinctions, which nowadays we would consider to be essentially the same except for the signs of the numbers. Omar distinguished *fourteen* different types of cubic, depending on which terms appear on each side of the equation. Omar's classification of cubic equations went like this:

cube = square + side + number

cube = square + number

cube = side + number

cube = number

cube + square = side + number

cube + square = number

cube + side = square + number

cube + side = number

cube + number = square + side

cube + number = square

cube + number = side

cube + square + side = number

cube + square + number = side

cube + side + number = square

Each listed term would have a positive numerical coefficient. You may be wondering why this list does not include cases like

cube + square = side

The reason is that in these cases we can divide both sides of the equation by the unknown, reducing it to a quadratic.

Omar did not entirely invent his solutions but instead built on earlier Greek methods for solving various types of cubic equation using conic sections. He developed these ideas systematically, and solved all fourteen types of cubic by such methods. Previous mathematicians, he noted, had discovered solutions of various cases, but these methods were all very special and each case was tackled by a different construction; no one before him had worked out the whole extent of possible cases, let alone found solutions to them. "Me, on the contrary—I have never ceased to wish to make known, with exactitude, all of the possible cases, and to distinguish among each of the cases the possible and impossible ones." By "impossible" he meant "having no *positive* solution."

To give a flavor of his work, here is how he solved "A cube, some sides, and some numbers are equal to some squares," which we would write as

$$x^3 + bx + c = ax^2.$$

(Since we don't care about positive versus negative, we would probably move the right-hand term to the other side and change a to $-a$ as well: $x^3 - ax^2 + bx + c = 0$).

Omar instructs his readers to carry out the following sequence of steps. (1) Draw three lines of lengths c/b, $\sqrt{b}$, and a, with a right angle. (2) Draw a semicircle whose diameter is the horizontal line. Extend the vertical line to cut it. If the solid vertical line has length d, make the solid horizontal line have length $cd/\sqrt{b}$. (3) Draw a hyperbola (solid line) whose asymptotes (those special straight lines that the curves approach) are the shaded lines,

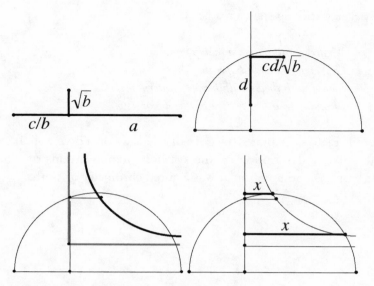

Omar Khayyám's solution of a cubic equation.

passing through the point just constructed. (4) Find where the hyperbola cuts the semicircle. Then the lengths of the two solid lines, marked *x,* are both (positive) solutions of the cubic.

The details, as usual, matter much less than the overall style. Carry out various Euclidean constructions with ruler and compass, throw in a hyperbola, carry out some more Euclidean constructions—done.

Omar gave similar constructions to solve each of his fourteen cases, and proved them correct. His analysis has a few gaps: the points required in his construction sometimes fail to exist when the sizes of the coefficients *a, b, c* are unsuitable. In the construction above, for example, the hyperbola may not meet the semicircle at all. But aside from these quibbles, he did an impressive and very systematic job.

Some of the imagery in Omar's poetry is mathematical and seems to allude to his own work, in the self-deprecatory tone that we find throughout:

> *For "Is" and "Is-Not," though with Rule and Line*
> *And "Up-And-Down" by logic I define,*
> *Of all that one should care to fathom, I*
> *Was never deep in anything but—Wine.*

One especially striking stanza reads:

> *We are no other than a moving row*
> *Of Magic Shadow-shapes that come and go*
> *Round with the Sun-illumined Lantern held*
> *In midnight by the Master of the Show.*

This recalls Plato's celebrated allegory of shadows on a cave wall. It serves equally well as a description of the symbolic manipulations of algebra, and the human condition. Omar was a gifted chronicler of both.

4

THE GAMBLING SCHOLAR

"I swear to you, by God's holy Gospels, and as a true man of honor, not only never to publish your discoveries, if you teach me them, but I also promise you, and I pledge my faith as a true Christian, to note them down in code, so that after my death no one will be able to understand them."

This solemn oath was—allegedly—sworn in 1539.

Renaissance Italy was a hotbed of innovation, and mathematics was no exception. In the iconoclastic spirit of the age, Renaissance mathematicians were determined to overcome the limitations of classical mathematics. One of them had solved the mysterious cubic. Now he was accusing another of stealing his secret.

The irate mathematician was Niccolo Fontana, nicknamed "Tartaglia," the stammerer. The alleged thief of his intellectual property was a mathematician, a doctor, an incorrigible rogue, and an inveterate gambler. His name was Girolamo Cardano, aka Jerome Cardan. Around 1520, Girolamo, a true prodigal son, had worked his way through his father's legacy. Broke, he turned to gambling as a source of finance, putting his mathematical abilities to effective use in assessing the chances of winning. He kept dubious company; once, when he suspected another player of cheating, he slashed the man's face with a knife.

They were hard times, and Girolamo was a hard man. He was also a highly original thinker, and he wrote one of the most famous and influential algebra texts in history.

We know a lot about Girolamo because in 1575 he told us all about himself in *The Book of My Life*. It begins thus:

> This Book of My Life I am undertaking to write after the example of Antoninus the Philosopher, acclaimed the wisest and best of men, knowing well that no accomplishment of mortal man is perfect, much less safe from calumny; yet aware that none, of all ends which man may attain, seems more pleasing, none more worthy than recognition of the truth.
>
> No word, I am ready to affirm, has been added to give savor of vainglory, or for sake of mere embellishment; rather, as far as possible, mere experiences were collected, events of which my pupils . . . had some knowledge, or in which they took part. These brief cross-sections of my history were in turn written down by me in narrative form to become this my book.

Like many mathematicians of the period, Girolamo practiced astrology, and he notes the astrological circumstances surrounding his birth:

> Although various abortive medicines—as I have heard—were tried in vain, I was normally born on the 24th day of September in the year 1500, when the first hour of the night was more than half run, but less than two thirds . . . Mars was casting an evil influence on each luminary because of the incompatibility of their positions, and its aspect was square to the moon.
>
> . . . I could easily have been a monster, except for the fact that the place of the preceding conjunction had been 29° in Virgo, over which Mercury is the ruler. And neither this planet nor the position of the moon or of the ascendant is the same, nor does it apply to the second decanate of Virgo; consequently I ought to have been a monster, and indeed was so near it that I came forth literally torn from my mother's womb.
>
> So I was born, or rather taken by violent means from my mother; I was almost dead. My hair was black and curly. I was revived in a bath of warm wine which might have been fatal to any other child. My mother had been in labor for three entire days, and yet I survived.

One chapter of *The Book of My Life* lists the books Girolamo wrote, and the first on the list is *The Great Art,* one of three "treatises in mathematics" that he mentions. He also wrote on astronomy, physics, morality, gemstones, water, medicine, divination, and theology.

Only *The Great Art* plays a part in our tale. Its subtitle, *The Rules of Algebra,* explains why. In it, Girolamo assembled methods for solving not just the quadratic equation, known to the Babylonians, but newly discovered solutions for cubic and quartic equations. Unlike Khayyám's solutions, which depended on the geometry of conics, those in *The Great Art* are purely algebraic.

Earlier, I mentioned two kinds of mathematical symbol, both of which we see in an expression such as x^3, for the cube of the unknown. The first kind of symbol is the use of letters (x in this case) to stand for numbers—either unknown, or known but arbitrary. The second kind uses raised numbers to indicate powers—so the superscript ^{3} here indicates the cube $x \times x \times x$. Now we come to a third kind of symbol, the last that we will need.

This third type of symbol is very pretty, and it looks like this: $\sqrt{}$. This symbol means "square root." For instance, $\sqrt{9}$, "the square root of nine," means the number that *when multiplied by itself* gives the answer 9. Since $3 \times 3 = 9$, we see that $\sqrt{9} = 3$. It's not always that easy, however. The most notorious square root, which according to a very unlikely legend caused the mathematician who drew attention to it, Hippasus of Metapontum, to be thrown overboard from a boat, is the square root of two, $\sqrt{2}$. An exact expression in decimals has to go on forever. It starts like this:

$$1.4142135623730950488,$$

but it can't *stop* there, because the square of that number is actually

$$1.99999999999999999999522356663907438144,$$

which obviously is not quite the same as 2.

This time we do know where the symbol came from. It is a distorted form of the letter "r," standing for "radix," the Latin for "root." Mathematicians understand it that way and read $\sqrt{2}$ as "root two."

Cube roots, fourth roots, fifth roots, and so on are shown by putting a small raised number in front of the "root" sign, like this:

$$\sqrt[3]{}, \sqrt[4]{}, \sqrt[5]{}.$$

The cube root of a given number is the number that has the given number as its cube, and so on. So the cube root of 8 is 2, because $2^3 = 8$. Again, the cube root of 2 can be expressed only approximately in decimal notation. It starts out like this:

$$1.2599210498948731648$$

and continues, if you have sufficient patience, forever.

It is this number that turns up in the ancient problem of doubling the cube.

By the year 400, Greek mathematics was no longer on the cutting edge. The action moved east, to Arabia, India, and China. Europe descended into the "Dark Ages," and while these were not quite as dark as they have often been painted, they were dark enough. The spread of Christianity had the unfortunate side effect of concentrating learning and scholarship in the churches and monasteries. Many monks copied the works of mathematical greats like Euclid, but very few of them understood what they were copying. The Greeks could dig a tunnel through a mountain from both ends and make it meet in the middle; the early Anglo-Saxon method of surveying was to lay out a design, *full scale,* in a field. Even the notion of drawing to scale had been lost. If the Anglo-Saxons had wanted to make an accurate map of England, they would have made it the same *size* as England. They did make maps of conventional size, but not very accurate ones.

By the end of the fifteenth century, the focus of mathematical activity was once again swinging back to Europe. As the Middle and Far East ran out of creative steam, Europe was getting its second wind, struggling free of the embrace of the Church of Rome and its fear of anything new. Ironically, the new center of intellectual activity was Italy, as Rome lost its grip on its own backyard.

This sea change in European science and mathematics began with the publication, in 1202, of a book called the *Liber Abbaci,* written by Leonardo of Pisa, who much later was given the nickname Fibonacci—son of Bonaccio—and is now known by that name even though it was invented in the nineteenth century. Leonardo's father, Guilielmo, was a customs officer in Bugia, now Algeria, and in his work must have come across people from many cultures. He taught his son the newfangled numerical symbols invented by the Hindus and the Arabs, the forerunners of our decimal digits 0 through 9. Leonardo later wrote that he "enjoyed so much the instruction that I continued to study mathematics while on business trips to Egypt, Syria, Greece, Sicily, and Provence, and there enjoyed disputations with the scholars of those places."

At first sight, the title of Leonardo's book seems to indicate that it is about the abacus, a mechanical calculating device using beads that slide on wires, or pebbles in a groove in the sand. But just as the Latin word calculus, referring to one of those small pebbles, later acquired a different and more technical meaning, so the word abbaco, the counting frame, came to mean the art of computation. The *Liber Abbaci* was the first arithmetic text to bring the Hindu-Arabic symbols and methods to Europe. A large part of it is given over to the new arithmetic's applications to practical subjects like currency exchange.

One problem, about an idealized model of the growth of a population of rabbits, led to the remarkable sequence of numbers 1, 1, 2, 3, 5, 8, 13, 21, 34, 55, and so on, where each number from 2 onward is the sum of the preceding two numbers. This "Fibonacci sequence" is Leonardo's greatest claim to fame—not for its rabbit-breeding implications, which are nil, but for its remarkable mathematical patterns and its key role in the theory of irrational numbers. Leonardo could have had no idea that this little *jeu d'esprit* would come to eclipse the entire rest of his life's work.

Leonardo wrote several other books, and his *Practica Geometriae* of 1220 contained a large part of Euclid, plus some Greek trigonometry. Book X of Euclid's *Elements* discusses irrational numbers composed of nested square roots, of the type $\sqrt{a+\sqrt{b}}$. Leonardo proved that these irrationals are inadequate for solving cubic equations. This does not imply that the roots of the cubic cannot be constructed by ruler and compass, because other combinations of square roots might yield solutions. But it was the first hint that cubics might not be solvable using only Euclid's tools.

In 1494, Luca Pacioli pulled together much existing mathematical knowledge in a book on arithmetic, geometry, and proportion. It included the Hindu-Arabic numerals, commercial mathematics, a summary of Euclid, and Ptolemy's trigonometry. A running theme was the element of design in nature, embodied in proportions—the human body, perspective in art, the theory of color.

Pacioli continued the tradition of "rhetorical" algebra, using words rather than symbols. The unknown was "thing," the Italian word *cosa,* and for a time, practitioners of algebra were known as "cossists." He also employed some standard abbreviations, continuing (but failing to improve on) the approach pioneered by Diophantus. Morris Kline makes a telling point in his monumental *Mathematical Thought from Ancient to Modern Times:* "It is a significant commentary on the mathematical development of arithmetic and algebra between 1200 and 1500 that Pacioli's [book] contained hardly anything more than Leonardo of Pisa's *Liber Abbaci.* In fact, the arithmetic and algebra . . . were based on Leonardo's book."

At the end of his book, Pacioli remarked that solving the cubic was no better understood than squaring the circle. But this would soon change. The first big breakthrough came about one-third of the way into the sixteenth century, in the city of Bologna. At first it passed unnoticed.

Girolamo Cardano was the bastard son of a Milan lawyer, Fazio Cardano, and a young widow named Chiara Micheria, the mother of three children by her former marriage. He was born in Pavia, a town in the duchy of Milan, in 1501. When the plague came to Milan, the pregnant Chiara was persuaded to move to the countryside, where she gave birth to Girolamo. Her three older children, who had remained behind, all died of the plague.

According to Girolamo's autobiography,

> my father went dressed in a purple cloak, a garment which was unusual in our community; he was never without a small black skullcap . . . From his fifty-fifth year on he lacked all of his teeth. He was well acquainted with the works of Euclid; indeed his shoulders were rounded from much study . . . My mother was easily provoked; she was quick of memory and wit, and a fat devout little woman. To be hasty-tempered was a trait common to both parents.

Though a lawyer by profession, Fazio was skilled enough in mathematics to give advice about geometry to Leonardo da Vinci. He taught geometry at the University of Pavia and at the Piatti Foundation, a Milanese institution. And he taught mathematics and astrology to his illegitimate son, Girolamo:

> My father, in my earliest childhood, taught me the rudiments of arithmetic, and about that time made me acquainted with the arcane; whence he had come by this learning I do not know. Shortly after, he instructed me in the elements of the astrology of Arabia . . . After I was twelve years old he taught me the first six books of Euclid.

The child had health problems; attempts to involve him in the family business were not successful. Girolamo managed to persuade his doubting father to let him study medicine at the University of Pavia. His father preferred law.

In 1494, Charles VIII of France had invaded Italy, and the ensuing war continued sporadically for fifty years. An outbreak of hostilities closed the University of Pavia, and Girolamo moved to Padua to continue his studies. By all accounts he was a first-class student, and when Fazio died, Girolamo was campaigning to become the university's rector. Although many people disliked him for speaking his mind, he was appointed by the margin of a single vote.

This is when he frittered away his inheritance and turned to gambling, which became an addiction for the rest of his turbulent life. And not only that:

> At a very early period in my life, I began to apply myself seriously to the practice of swordsmanship of every class, until, by persistent training, I had acquired some standing even among the most daring . . . By night, even contrary to the decrees of the Duke, I armed myself and went prowling about the cities in which I dwelt . . . I wore a black woolen hood to conceal my features, and put on shoes of sheep-pelt . . . often I wandered abroad throughout the night until day broke, dripping with perspiration from the exertion of serenading on my musical instruments.

It scarcely bears thinking about.

Awarded his medical degree in 1525, Girolamo tried to enter the College of Physicians in Milan and was rejected—nominally for illegitimacy but in fact, largely because of his notorious lack of tact. So instead of joining the prestigious college, Girolamo set himself up as a doctor in the nearby village of Sacco. This provided a small income, but the business limped along. He married Lucia Bandarini, the daughter of a captain in the militia, and moved closer to Milan, hoping to earn more money to provide for his family, but again the college turned him down. Unable to pursue a legitimate medical career, he reverted to gambling, but even his mathematical expertise failed to restore his fortunes:

> Peradventure in no respect can I be deemed worthy of praise; for so surely as I was inordinately addicted to the chess-board and the dicing table, I know that I must rather be considered deserving of the severest censure. I gambled at both for many years, at chess more than forty years, at dice about twenty-five; and not only every year, but—I say it with shame—every day, and with the loss at once of thought, of substance, and of time.

The entire family ended up in the poor house, having long ago pawned their furniture and Lucia's jewelry. "I entered upon a long and honorable career. But away with honors and gain, together with vain displays and unseasonable delights! I ruined myself! I perished!"

Their first child arrived:

> After having twice miscarried and borne two males of four months, so that I . . . at times suspected some malefic influence, my wife brought forth my first born son . . . He was deaf in his right ear . . . Two toes on his left foot . . . were joined by one membrane. His back was slightly hunched but not to the extent of a deformity. The boy led a tranquil existence up to his twenty-third year. After that, he fell in love . . . and married a dowerless wife, Brandonia di Seroni.

Now Girolamo's late father came to their rescue, in a rather indirect manner. Fazio's lecturing post at the university was still open, and Girolamo got the job. He also did a bit of doctoring on the side, despite being unlicensed. A number of miraculous cures—probably luck, given the state of medicine in that period—gave him a high reputation. Even some members of the college took their medical problems to him, and for a while it looked as though he might finally gain entrance to that esteemed institution. But once again, Girolamo's tendency to speak his mind scuppered that; he published a vitriolic attack on the abilities and character of the college's membership. Girolamo was aware of his lack of tact but apparently did not see it as a fault: "As a lecturer and debater, I was much more earnest and accurate than in exercising prudence." In 1537 his lack of prudence caused his latest application to be turned down.

But his reputation was becoming so great that the college eventually had no real choice, and he was made a member two years later. Things were looking up; all the more so when he published two books about mathematics. Girolamo's career was advancing on several fronts.

Around this time, Tartaglia made a major breakthrough—a solution to a broad class of cubic equations. After some persuasion, and with reluctance, he confided his epic discovery to Cardano. It is hardly surprising that six years later, when he received a copy of Cardano's algebra text *The Great Art, or on the Rules of Algebra* and found a complete exposition of his secret, Tartaglia was incensed.

Cardano had not stolen the *credit,* for he gave full acknowledgment to Tartaglia:

> In our own days Scipione del Ferro of Bologna has solved the case of the cube and first power equal to a constant, a very elegant and admirable accomplishment . . . In emulation of him, my friend Niccolò Tartaglia of Brescia . . . solved the same case when he got into a contest with his [del Ferro's] pupil Antonio Maria Fior, and moved by my many entreaties, gave it to me.

Nonetheless, it was galling for Tartaglia to see his precious secret given away to the world, and even more galling to recognize that many more people would remember the author of the book than the erstwhile secret's true discoverer.

That, at least, was Tartaglia's view of the affair, which constitutes almost all of the existing evidence. As Richard Witmer points out in his translation of *The Great Art,* "We are dependent almost exclusively on Tartaglia's printed accounts, which by no stretch of the imagination can be regarded as objective." One of Cardano's servants, Lodovico Ferrari, later claimed to have been present at the meeting and said that there had been no agreement to keep the method secret. Ferrari later became Cardano's student, and he solved—or helped to solve—the quartic, so he cannot be considered a more objective witness than Tartaglia.

What made matters worse for poor Tartaglia was that it wasn't just a case of lost credit. In Renaissance Europe, mathematical secrets could be translated into hard cash. Not just through gambling, Cardano's preferred route, but through public competitions.

It is often said that mathematics is not a spectator sport, but that was not true in the 1500s. Mathematicians made reasonable livings by challenging each other to public contests, in which each would set his opponent a series of problems, and whoever got the most answers right was the winner. These spectacles were less thrilling than bare-hands fighting or swordplay, but onlookers could place wagers and find out which contestant won, even if they had no idea how he did it. In addition to the prize money, winners attracted pupils, who would pay for tuition, so the public competitions were doubly lucrative.

Tartaglia was not the first to find an algebraic solution to a cubic equation. The Bolognese professor Scipione del Ferro discovered his solution of some types of cubic somewhere around 1515. He died in 1526, and both his papers and his professorship were inherited by his son-in-law, Annibale del Nave. We can be sure of this because the papers themselves came to light in the University of Bologna library around 1970, thanks to the efforts of E. Bartolotti. According to Bartolotti, del Ferro probably knew how to solve three types of cubic, but he passed on the method for solving only one type: cube plus thing equals number.

Knowledge of this solution was preserved by del Nave and by del Ferro's student Antonio Maria Fior. And it was Fior, determined to set himself up in business as a mathematics teacher, who came up with an effective marketing technique. In 1535 he challenged Tartaglia to a public cubic-solving contest.

There were rumors that a method for solving cubics had been found, and nothing encourages a mathematician more than the knowledge that a problem *has* a solution. The risk of wasting time on an unsolvable problem is ruled out; the main danger is that you may not be clever enough to find an answer you know must exist. All you need is lots of confidence, which mathematicians seldom lack—even if it turns out to be misplaced.

Tartaglia had rediscovered del Ferro's method, but he suspected that Fior also knew how to solve other types of cubic and would thus have a huge advantage. Tartaglia tells us how much this prospect worried him, and how he finally cracked the remaining case shortly before the contest. Now Tartaglia had the advantage, and he promptly wiped out the unlucky Fior.

Word of the defeat spread; Cardano heard of it in Milan. At that time he was working on his algebra text. Like any true author, Cardano was determined to include the latest discoveries, for without them his book would be obsolete before it was even published. So Cardano approached Tartaglia, hoping to wheedle the secret out of him and put it in *The Great Art*. Tartaglia refused, saying that he intended to write his own book.

Eventually, however, Cardano's persistence paid off, and Tartaglia divulged the secret. Did he *really* make Cardano swear to keep it hidden, knowing that a textbook was in the offing? Or did he succumb to Cardano's blandishments and then regret it?

There is no doubt that he was extremely angry when *The Great Art* appeared. Within a year he had published a book, *Diverse Questions and Inventions,* which laid into Cardano in no uncertain terms. He included all of the correspondence between them, supposedly exactly as written.

In 1574, Ferrari came to the support of his master by issuing a *cartello*—a challenge to a learned dispute on any topic Tartaglia cared to name. He even offered a prize of 200 scudi for the winner. And he made his opinions very clear: "This I have proposed to make known, that you have written things which falsely and unworthily slander . . . Signor Girolamo, compared to whom you are hardly worth mentioning."

Ferrari sent copies of the *cartello* to numerous Italian scholars and public figures. Within nine days, Tartaglia responded with his own statement of facts, and the two mathematicians ended up by exchanging twelve *cartelli* between them over a period of eighteen months. The dispute seems to have followed the standard rules for a genuine duel. Tartaglia, who had been insulted by Ferrari, was allowed the choice of weapons— the selection of topics to be debated. But he kept asking to debate Cardano rather than his challenger, Ferrari.

Ferrari kept his temper under control and pointed out that in any case it had been del Ferro, not Tartaglia, who had solved the cubic to begin with.

Since del Ferro had made no fuss about Tartaglia's unjustified claim of credit, why wasn't Tartaglia willing to behave likewise? It was a good point, and Tartaglia may have recognized that, because he considered withdrawing from the contest. However, he didn't, and one possible reason was the city fathers of Brescia, his hometown. Tartaglia was after a lectureship there, and the local dignitaries may have wanted to see how he acquitted himself.

At any rate, Tartaglia agreed to the debate, which was held in a Milanese church before large crowds in August 1548. No record of the proceedings is known, save for a few indications by Tartaglia, who said that the meeting ceased when suppertime approached. This hints that the debate may not have been especially gripping. It seems, though, that Ferrari won handily, because afterward he was offered some plum positions, accepted the post of tax assessor to the governor of Milan, and soon became very rich. Tartaglia, on the other hand, never claimed to have won the debate, failed to get the Brescia job, and descended into bitter recrimination.

Unknown to Tartaglia, Cardano and Ferrari had an entirely different line of defense, for they had visited Bologna and inspected del Ferro's papers. These included the first genuine solution of the cubic, and in later years they both insisted that the source of the material included in *The Great Art* was del Ferro's original writings, not Tartaglia's confidence to Cardano. The reference to Tartaglia was included merely to record how Cardano himself heard of del Ferro's work.

There is a final twist to the tale. Soon after the second edition of *The Great Art* was published, in 1570, Cardano was imprisoned by the Inquisition. The reason may have been something that had previously seemed entirely innocent: not the content of the book, but its dedication. Cardano had chosen to dedicate it to the relatively obscure intellectual Andreas Osiander, a minor figure in the Reformation but one strongly suspected of being the author of an anonymous preface to Nicolaus Copernicus's *On the Revolutions of the Heavenly Spheres,* the first book to propose that the planets go around the Sun, not the Earth. The Church considered this view heretical, and in 1600 it burned Giordano Bruno alive for maintaining it—hanging him upside down from a stake, naked and gagged, in a Roman market square. In 1616, and again in 1633, it gave Galileo a lot of grief, for the same reason, but by then the Inquisition was content to put him under house arrest.

✳

To appreciate what Girolamo and his compatriots achieved, we must re-visit the Babylonian tablet that explains how to solve quadratics. If we follow the recipe and express the calculation steps in modern symbolism, we see that in effect the Babylonian scribe was saying that the solution to a quadratic equation $x^2 - ax = b$ is

$$x = \sqrt{\left(\tfrac{a}{2}\right)^2 + b} + \tfrac{a}{2}$$

This is equivalent to the formula that every school student used to learn by heart, and that nowadays is found in every formula book.

The Renaissance solution of the cubic equation is similar but more elaborate. In modern symbols it looks like this: Suppose that $x^3 + ax = b$. Then

$$x = \sqrt[3]{\frac{b}{2} + \sqrt{\frac{a^3}{27} + \frac{b^2}{4}}} + \sqrt[3]{\frac{b}{2} - \sqrt{\frac{a^3}{27} + \frac{b^2}{4}}}$$

As formulas go, this one is relatively simple (believe me!), but you need to develop a lot of algebraic ideas before it can be so described. It is by far the most complicated formula we will look at, and it uses all three types of symbol that I have introduced: letters, raised numbers, and the $\sqrt{}$ sign, as well as both square roots and cube roots. You don't need to understand this formula, and you certainly don't need to calculate it. But you need to appreciate its general shape. First, some terminology that will prove very useful as we proceed.

An algebraic expression like $2x^4 - 7x^3 - 4x^2 + 9$ is called a *polynomial,* which means "many terms." Such expressions are formed by adding various powers of the unknown. The numbers 2, −7, −4, and 9, which multiply the powers, are called *coefficients.* The highest power of the unknown that occurs is called the *degree* of the polynomial, so this polynomial has degree 4. There are special names for polynomials of low degree (1 through 6): linear, quadratic, cubic, quartic, quintic, and sextic. The solutions of the associated equation $2x^4 - 7x^3 - 4x^2 + 9 = 0$ are called the *roots* of the polynomial.

Now we can dissect Cardano's formula. It is built from the coefficients *a* and *b,* employing some additions, subtractions, multiplications, and divisions (but only by certain integers, namely 2, 4, and 27). The esoteric aspects are of two kinds: There is a square root—in fact the same square root occurs in two places, but one is added while the other is subtracted.

Finally, there are two cube roots, and these are the cube roots of quantities that involve the square roots. So aside from harmless operations of algebra (by which I mean those that merely shuffle the terms around), the skeleton of the solution is, "Take a square root, then a cube root; do this again; add the two."

That's all we will need. But I don't think we can get away with less.

What the Renaissance mathematicians initially failed to grasp, but later generations soon realized, is that this formula is not just a solution to one type of cubic. It is a complete solution to *all* types, give or take some straightforward algebra. For a start, if the cube term is, say, $5x^3$ instead of x^3, you can just divide the entire equation by 5—the Renaissance mathematicians were certainly smart enough to spot that. A more subtle idea, which required a quiet revolution in how we think of numbers, is that by allowing the coefficients a and b to be *negative* if necessary, we can avoid fruitless distinctions among cases. Finally, there is a purely algebraic trick: if the equation involves the square of the unknown, you can always get rid of it—you replace x by x plus a carefully chosen constant, and if you do it right, the square term miraculously disappears. Again, it helps here if you stop worrying whether numbers are positive or negative. Finally, the Renaissance mathematicians worried about terms that were entirely missing, but to modern eyes the remedy is obvious: such a term is not actually *missing*, it just has coefficient zero. The same formula works.

Problem solved?

Not quite. I lied.

Here is where I lied. I said that Cardano's formula solves *all* cubics. There is a sense in which that's not true, and it turned out to be important. I didn't tell a very bad lie, though, because it all depends on what you mean by "solve."

Cardano himself spotted the difficulty, which says a lot for his attention to detail. Cubics typically have either three solutions (fewer if we exclude negative numbers) or one. Cardano noticed that when there are three solutions—say 1, 2, and 3—the formula does not seem to yield those solutions in any sensible way. Instead, the square root in the formula contains a negative number.

Specifically, Cardano noted that the cubic $x^3 = 15x + 4$ has the obvious solution $x = 4$. But when he tried out Tartaglia's formula, it led to the "answer"

$$x = \sqrt[3]{2 + \sqrt{-121}} + \sqrt[3]{2 - \sqrt{-121}}$$

which seemed meaningless.

Among European mathematicians in those days, few brave souls were willing to contemplate negative numbers. Their Eastern counterparts had come to terms with negative quantities much earlier. In India, the Jains developed a rudimentary concept of negative quantities as early as 400, and in 1200 the Chinese system of "counting rods" used red rods for positive numbers and black rods for negative ones—though only in certain limited contexts.

If negative numbers were a puzzle, their square roots had to be even more baffling. The difficulty is that the square of either a positive or a negative number is always *positive*—I won't explain why here, but it is the only choice that makes the laws of algebra work consistently. So even if you are happy using negative numbers, it seems that you have to accept that they cannot have sensible square roots. Any algebraic expression involving the square root of a negative quantity must therefore be nonsense.

And yet Tartaglia's formula led Cardano to just such an expression. It was worrying in the extreme that in cases in which you *knew* the solution by some other route, the formula seemed not to produce it.

In 1539 a worried Cardano raised the matter with Tartaglia:

I have sent to enquire after the solution to various problems for which you have given me no answer, one of which concerns the cube equal to an unknown plus a number. I have certainly grasped this rule, but when the cube of one-third of the coefficient of the unknown is greater in value than the square of one-half of the number, then, it appears, I cannot make it fit into the equation.

Here Cardano is describing exactly the condition for the square root to be that of a negative number. It is clear that he had an excellent grasp of the whole business and had spotted a snag. It is less clear whether Tartaglia had a comparable level of understanding of his own formula, because his response was, "you have not mastered the true way of solving problems of this kind . . . Your methods are totally false."

Possibly Tartaglia was merely being deliberately unhelpful. Or possibly he did not see what Cardano was getting at. At any rate, Cardano had put his finger on a puzzle that would exercise the combined intellects of the world's mathematicians for the next 250 years.

Even in Renaissance times, there were hints that something important might be going on. The same issue arose in another problem discussed in *The Great Art,* to find two numbers whose sum is 10 and whose product is 40. This led to the "solution" $5 + \sqrt{-15}$ and $5 - \sqrt{-15}$. Cardano noticed that if you ignored the question of what the square root of minus fifteen *meant,* and just pretended it worked like any other square root, then you could check that these "numbers" actually fit the equation. When you added them, the square roots canceled out, and the two 5's added to 10, as required. When you multiplied them, you got $5^2 - (\sqrt{-15})^2$, which equals 25 + 15, which is 40. Cardano did not know what to make of this strange calculation. "So," he wrote, "progresses arithmetic subtlety, the end of which is as refined as it is useless."

In his *Algebra* of 1572, Rafaele Bombelli, the son of a Bolognese wool merchant, noticed that similar calculations, manipulating the "imaginary" roots as if they were genuine numbers, could convert the weird formula for Cardano's puzzling cubic into the correct answer, $x = 4$. He wrote the book to fill in some spare time while he was reclaiming marshland for the Apostolic Camera, the Pope's legal and financial department. Bombelli noticed that

$$(2 + \sqrt{-1})^3 = 2 + \sqrt{-121}$$

and

$$(2 - \sqrt{-1})^3 = 2 - \sqrt{-121}$$

so the sum of the two strange cube roots becomes

$$(2 + \sqrt{-1}) + (2 - \sqrt{-1})$$

which equals 4. The meaningless root was somehow meaningful, and it gave the right answer. Bombelli was probably the first mathematician to realize that you could carry out algebraic manipulations with square roots of negative numbers and get usable results. This was a big hint that such numbers had a sensible interpretation, but it didn't seem to indicate what that interpretation was.

The mathematical high point of Cardano's book was not the cubic but the quartic. His student Ferrari managed to extend Tartaglia and del Ferro's methods to equations that contain the fourth power of the unknown. Ferrari's formula involves only square roots and cube roots—a fourth root is just the square root of a square root, so those are not needed.

The Great Art does not include a solution of the quintic equation, in which the unknown appears to the fifth power. But as the degree of the equation increases, the method for solving it gets more and more complicated, so few doubted that with enough ingenuity, the quintic too could be solved—you probably had to use fifth roots, and any formula would be really messy.

Cardano did not spend time seeking such a solution. After 1539 he returned to his numerous other activities, especially medicine. And now his family life fell apart in the most horrific manner: "My [youngest] son, between the day of his marriage and the day of his doom, had been accused of attempting to poison his wife while she was still in the weakness attendant upon childbirth. On the 17th day of February he was apprehended, and fifty-three days after, on April 13th, he was beheaded in prison." While Cardano was still trying to come to terms with that tragedy, the horror got worse. "One house—mine—witnessed within the space of a few days, three funerals, that of my son, of my little granddaughter, Diaregina, and of the baby's nurse; nor was the infant grandson far from dying."

For all that, Cardano was incurably optimistic about the human condition: "Nevertheless, I still have so many blessings, that if they were another's he would count himself lucky."

5

THE CUNNING FOX

hich road to take? Which subject to study? He loved them both, but he must choose between them. It was a terrible dilemma. The year was 1796, and a brilliant 19-year-old youth was faced with a decision that would affect the rest of his life. He had to decide on a career. Although he came from an ordinary family, Carl Friedrich Gauss knew that he could rise to greatness. Everyone recognized his ability, including the duke of Brunswick, in whose domain Gauss had been born and where his family lived. His problem was that he had too much ability, and he was forced to choose between his two great loves: mathematics and linguistics.

On 30 March, however, the decision was taken out of his hands by a curious, remarkable, and totally unprecedented discovery. On that day, Gauss found a Euclidean construction for a regular polygon with seventeen sides.

This may sound esoteric, but there was not even a hint of it in Euclid. You could find methods for constructing regular polygons with three sides, or four, or five, or six. You could combine the constructions for three and five sides to get fifteen, and repeated bisections would double the number of sides, leading to eight, ten, twelve, sixteen, twenty, . . .

But seventeen was crazy. It was also true, and Gauss knew full well *why* it was true. It all boiled down to two simple properties of the number 17. It is a prime number—its only exact divisors are itself and 1. And it is one greater than a power of two: $17 = 16 + 1 = 2^4 + 1$.

If you were a genius like Gauss, you could see why those two unassuming statements implied the existence of a construction, using straightedge and compass, of the regular seventeen-sided polygon. If you were any of the other great mathematicians who had lived between 500 BCE and 1796, you would not even have got a sniff of any connection. We know this because they didn't.

If Gauss had needed confirmation of his mathematical talent, he certainly had it now. He resolved to become a mathematician.

The Gauss family had moved to Brunswick in 1740 when Carl's grandfather took a job there as a gardener. One of his three sons, Gebhard Dietrich Gauss, also became a gardener, occasionally working at other laboring jobs such as laying bricks and tending canals; at other times he was a "master of waterworks," a merchant's assistant, and the treasurer of a small insurance fund. The more profitable trades were all controlled by guilds, and newcomers—even second-generation newcomers—were denied access to them. Gebhard married his second wife, Dorothea Benze, a stonemason's daughter working as a maid, in 1776. Their son Johann Friederich Carl (who later always called himself Carl Friedrich) was born in 1777.

Gebhard was honest but obstinate, ill-mannered, and not very bright. Dorothea was intelligent and self-assertive, traits that worked to Carl's advantage. By the time the boy was two, his mother knew she had a prodigy on her hands, and she set her heart on ensuring that he received an education that would allow his talents to flourish. Gebhard would have been happier if Carl had become a bricklayer. Thanks to his mother, Carl rose to fulfill the prediction that his friend, the geometer Wolfgang Bolyai, made to Dorothea when her son was 19, saying that Carl would become "the greatest mathematician in Europe." She was so overjoyed that she burst into tears.

The boy responded to his mother's devotion, and for the last two decades of her life she lived with him, her eyesight failing until she became totally blind. The eminent mathematician insisted on looking after her himself, and he nursed her until 1839, when she died.

Gauss showed his talents early. At the age of three, he was watching his father, at that point a foreman in charge of a gang of laborers, handing out the weekly wages. Noticing a mistake in the arithmetic, the boy pointed it out to the amazed Gebhard. No one had taught the child numbers. He had taught himself.

A few years later, a schoolmaster named J. G. Büttner set Gauss's class a task that was intended to occupy them for a good few hours, giving the teacher a well-earned rest. We don't know the exact question, but it was something very similar to this: add up all of the numbers from 1 to 100. Most likely, the numbers were not as nice as that, but there was a hidden pattern to them: they formed an arithmetic progression, meaning that the difference between any two consecutive numbers was always the same. There is a simple but not particularly obvious trick for adding the numbers in an arithmetic progression, but the class had not been taught it, so they had to laboriously add the numbers one at a time.

At least, that's what Büttner expected. He instructed his pupils that as soon as they had finished the assignment, they should place their slate, with the answer, on his desk. While his fellow students sat scribbling things like

$$1 + 2 = 3$$

$$3 + 3 = 6$$

$$6 + 4 = 10$$

with the inevitable mistake

$$10 + 5 = 14$$

and running out of space to write in, Gauss thought for a moment, chalked one number on his slate, walked up to the teacher, and slapped the slate face down on the desk.

"There it lies," he said, went back to his desk, and sat down.

At the end of the lesson, when the teacher collected all the slates, precisely one had the correct answer: Gauss's.

Again, we don't know exactly how Gauss's mind worked, but we can come up with a plausible reconstruction. In all likelihood, Gauss had already thought about sums of that kind and spotted a useful trick. (If not, he was entirely capable of inventing one on the spot.) An easy way to find the answer is to group the numbers in pairs: 1 and 100, 2 and 99, 3 and 98, and so on, all the way to 50 and 51. Every number from 1 to 100 occurs exactly once in some pair, so the sum of all those numbers is the sum of all the pairs. But each pair adds up to 101. And there are 50 pairs. So the total is $50 \times 101 = 5050$. This (or some equivalent) is what he chalked on his slate.

The point of this tale is not that Gauss was unusually good at arithmetic, though he was; in his later astronomical work he routinely carried out enormous calculations to many decimal places, working with the speed of an idiot savant. But lighting calculation was not his sole talent. What he possessed in abundance was a gift for spotting cryptic patterns in mathematical problems, and using them to find solutions.

Büttner was so astonished that Gauss had seen through his clever ploy that, to his credit, he gave the boy the best arithmetic textbook that money could buy. Within a week, Gauss had gone beyond anything his teacher could handle.

It so happened that Büttner had a 17-year-old assistant, Johann Bartels, whose official duties were to cut quills for writing and to help the boys use them. Unofficially, Bartels had a fascination for mathematics. He was drawn to the brilliant ten-year old, and the two became lifelong friends. They worked on mathematics together, each encouraging the other.

Bartels was on familiar terms with some of the leading lights of Brunswick, and they soon learned that there was an unsung genius in their midst, whose family lived on the brink of poverty. One of these gentlemen, councilor and professor E. A. W. Zimmerman, introduced Gauss to the duke of Brunswick, Carl Wilhelm Ferdinand, in 1791. The duke, charmed and impressed, took it upon himself to pay for Gauss's education, as he occasionally did for the talented sons of the poor.

Mathematics was not the boy's sole talent. By the age of 15 he had become proficient in classical languages, so the duke financed studies in classics at the local gymnasium. (In the old German educational system, a gymnasium was a type of school that prepared its pupils for university entrance. It translates roughly as "high school," but only paying students were admitted.) Many of Gauss's best works were later written in Latin. In 1792, he entered the Collegium Carolinium in Brunswick, again at the duke's expense.

By the age of 17 he had already discovered an astonishing theorem known as the "law of quadratic reciprocity" in the theory of numbers. It is a basic but rather esoteric regularity in divisibility properties of perfect squares. The pattern had already been noticed by Leonhard Euler, but Gauss was unaware of this and made the discovery entirely on his own. Very few people would even have thought of asking the question. And the boy was thinking very deeply about the theory of equations. In fact, that was what led him to his construction of the regular 17-gon and thus set him on the road to mathematical immortality.

Between 1795 and 1798, Gauss studied for a degree at the University of Göttingen, once more paid for by Ferdinand. He made few friends, but the friendships he did strike up were deep and long-lasting. It was at Göttingen that Gauss met Bolyai, an accomplished geometer in the Euclidean tradition.

Mathematical ideas came so thick and fast to Gauss that sometimes they seemed to overwhelm him. He would suddenly cease whatever he was doing and stare blankly into the middle distance as a new thought struck him. At one point he worked out some of the theorems that would hold "if Euclidean geometry were not the true one." At the forefront of his thoughts was a major work that he was composing, the *Disquisitiones Arithmeticae,* and by 1798 it was pretty much finished. But Gauss wanted to make certain that he had given due credit to his predecessors, so he visited the University of Helmstedt, which had a high-quality mathematics library overseen by Johann Pfaff, the best-known mathematician in Germany.

In 1801, after a frustrating delay at the printer's, the *Disquisitiones Arithmeticae* was published, with an effusive and no doubt heartfelt dedication to Duke Ferdinand. The duke's generosity did not end when Carl left the university. Ferdinand paid for his doctoral thesis, which he presented at the University of Helmstedt, to be printed as the regulations required. And when Carl started to worry about how to support himself when he left university, the duke gave him an allowance so that he could continue his researches without having to be bothered about money.

A noteworthy feature of the *Disquisitiones Arithmeticae* is its uncompromising style. The proofs are careful and logical, but the writing makes no concessions to the reader and gives no clue about the intuition behind the theorems. Later, he justified this attitude, which continued throughout his career, on the grounds that "When one has constructed a fine building, the scaffolding should no longer be visible." Which is all very well if all you want people to do is admire the building. It's not so helpful if you want to teach them how to build their own. Carl Gustav Jacob Jacobi, whose work in complex analysis built on Gauss's ideas, said of his illustrious predecessor, "He is like the fox, who erases his tracks in the sand with his tail."

Around this time, mathematicians were gradually coming to realize that although "complex" numbers seemed artificial and their meaning incomprehensible, they made algebra much neater by providing solutions to equations in a uniform way. Elegance and simplicity are the touchstones of mathematics, and novel concepts, however strange they appear at first, tend to win out in the long run if they help to *keep* the subject elegant and simple.

If you work purely with traditional "real" numbers, equations can be annoyingly erratic. The equation $x^2 - 2 = 0$ has two solutions, plus or minus the square root of two, but the very similar equation $x^2 + 1 = 0$ has none at all. However, this equation has two solutions in complex numbers: i and $-i$. The symbol i for $\sqrt{-1}$ was introduced by Euler in 1777 but not published until 1794. A theory couched solely in terms of "real" equations is littered with exceptions and pedantic distinctions. The analogous theory of complex equations avoids all of these complications by swallowing wholesale one big complication at the very outset: to allow complex numbers as well as real ones.

By 1750, the circle of ideas initiated by the mathematicians of Renaissance Italy had matured and closed. Their methods for solving cubic and quartic equations were seen as natural extensions of the Babylonian solution of quadratics. The connection between radicals and complex numbers had been worked out in some detail, and it was known that in this extension of the usual number system, a number had not one cube root but three; not one fourth root but four; not one fifth root but five. The key to understanding where these new roots came from was a beautiful property of "roots of unity," that is, nth roots of the number 1. These roots form the vertices of a regular n-sided polygon in the complex plane, with one vertex at 1. The remaining roots of unity space themselves out equally around a circle of radius 1, centered at 0. For instance, the left-hand figure (next page) shows the locations of the fifth roots of unity.

More generally, from any particular fifth root of some number it is possible to obtain four more, by multiplying it by q, q^2, q^3, and q^4. These numbers are also spaced equally around a circle centered at 0. For example, the five fifth roots of 2 are shown in the right-hand figure.

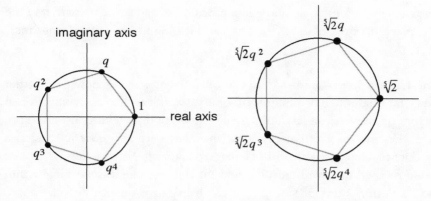

(Left) The fifth roots of unity in the complex plane. (Right) The fifth roots of two.

This was all very pretty, but it suggested something much deeper. The fifth roots of 2 can be viewed as the solutions of the equation $x^5 = 2$. This equation is of the fifth degree, and it has five complex solutions, only one of which is real. Similarly, the equation $x^4 = 2$ for fourth roots of 2 has four solutions, the equation for 17th roots of 2 has 17 solutions, and so on. You don't have to be a genius to spot the pattern: the number of solutions is equal to the degree of the equation.

The same seemed to apply not just to the equations for nth roots, but to *any algebraic equation whatsoever.* Mathematicians became convinced that within the realm of complex numbers, every equation has exactly the same number of solutions as its degree. (Technically, this statement is true only when solutions are counted according to their "multiplicity." If this convention is not used, then the number of solutions is less than or equal to the degree.) Euler proved this property for equations of degree 2, 3, and 4, and claimed that similar methods would work in general. His ideas were plausible, but filling in the gaps turned out to be almost impossible, and even today it takes a major effort to push Euler's method to a conclusion. Nevertheless, mathematicians assumed that if they were solving an equation of some degree, they should expect to find precisely that many solutions.

As Gauss developed his ideas in number theory and analysis, he became more and more dissatisfied that no one had proved this assumption. Characteristically, he came up with a proof. It was complicated and curiously indirect: any competent mathematician could be convinced that it was correct, but no one could guess how Gauss had come up with it in the first place. The fox of mathematics was wielding his tail with a vengeance.

The Latin title of Gauss's dissertation translates as "A new proof that every rational integral function of one variable can be resolved into real factors of the first or second degree." Unwrapping the jargon of the period, the title asserts that every polynomial (with real coefficients) is a product of terms that are either linear or quadratic polynomials.

Gauss used the word "real" to make it clear that he was working within the traditional number system, in which negative quantities lack square roots. Nowadays we would state Gauss's theorem in a logically equivalent but simpler form: every real polynomial of degree n has n real or complex roots. But Gauss chose his terminology carefully, so that his work did not rely on the still puzzling system of complex numbers. Complex roots of a real polynomial can always be combined in pairs to yield real quadratic factors, whereas linear factors correspond to real roots. By phrasing the title in terms of these two types of factor ("factors of the first or second degree"), Gauss circumvented the contentious issue of complex numbers.

One word in the title was unjustified: "new," which implies that there are "old" proofs. Gauss gave the *first* rigorous proof of this basic theorem in algebra. But to avoid offending illustrious predecessors who had already claimed proofs—all of them faulty—Gauss presented his breakthrough as merely the latest proof, using new (that is, correct) methods.

This theorem came to be known as the Fundamental Theorem of Algebra. Gauss considered it so important that he gave four proofs in all, the last at the age of 70. He personally had no qualms about complex numbers: they played a big role in his thinking, and he subsequently developed his own explanation of their meaning. But he disliked controversy. Over the years he suppressed many of his most original ideas—non-Euclidean geometry, complex analysis, and a rigorous approach to complex numbers themselves—because he did not want to attract what he referred to as "the cries of the Boeotians."

Gauss did not confine himself to pure mathematics. Early in 1801, the Italian priest and astronomer Giuseppe Piazzi had discovered a new planet, or so he thought—a faint patch of light in his telescope that moved against the background of the stars from one night to the next, a sure sign that it was a body in the solar system. It was duly given the name Ceres, but it was actually an asteroid, the first to be found. Having found the new world, Piazzi promptly lost it in the glare of the Sun. He had made so few observations that astronomers hadn't been able to work out the new body's orbit and worried that they wouldn't be able to locate it again when it emerged from behind the Sun.

This was a problem worthy of Gauss, and he set to with a will. He invented better ways to determine orbits from small numbers of observations, and predicted where Ceres would reappear. When it duly did so, Gauss's fame spread far and wide. The explorer Alexander von Humboldt asked Pierre-Simon de Laplace, an expert in celestial mechanics, to name the greatest mathematician in Germany, and got the reply "Pfaff." When a startled Humboldt asked, "What about Gauss?" Laplace replied, "Gauss is the greatest mathematician in the world."

Unfortunately, this newfound celebrity diverted him from pure mathematics into lengthy calculations in celestial mechanics—generally felt to be a waste of his considerable talents. It's not that celestial mechanics was unimportant, but other, less able mathematicians could have done the same work. On the other hand, it also set him up for life. Gauss had been looking for a prominent position that offered the opportunity for public service, to reward his sponsor, the duke. His work on Ceres landed him the directorship of the Göttingen observatory, a post that he held for the rest of his academic life.

He married Johanna Osthoff in 1805. Writing to Bolyai, he described his new wife: "The beautiful face of a Madonna, a mirror of peace of mind and health, tender, somewhat fanciful eyes, a blameless figure—this is one thing; a bright mind and an educated language—this is another; but the quiet, serene, modest and chaste soul of an angel who can do no harm to any creature—that is the best." Johanna bore him two children, but in 1809 she died in childbirth, and a devastated Gauss "closed her angelic eyes in which I have found heaven for the last five years." He became lonely and depressed, and life for him was never quite the same. He did find a new wife, Johanna's best friend Minna Waldeck, but the marriage was not terribly happy despite the birth of three more children. Gauss was always arguing with his sons and telling his daughters what to do, and the boys got so fed up that they left Europe for the United States, where they prospered.

Soon after taking up the directorship at Göttingen, Gauss returned to an old idea, the possibility of a new type of geometry that satisfies all of Euclid's axioms except the parallel axiom. He eventually became convinced that logically consistent non-Euclidean geometries are possible, but never published his results for fear that they would be considered too radical. János Bolyai, the son of his old friend Wolfgang, later made similar discoveries, but Gauss felt unable to praise the work because he had anticipated most of it. Still later, when Nikolai Ivanovich Lobachevsky independently rediscovered non-Euclidean geometry, Gauss had him made a corresponding member of the Göttingen Academy, but again offered no public praise.

Years later, as mathematicians studied these new geometries in more detail, they came to be interpreted as geometries of "geodesics"—shortest paths—on curved surfaces. If the surface had constant positive curvature, like a sphere, the geometry was called elliptic. If the curvature was constant and negative (shaped like a saddle near any point) the geometry was hyperbolic. Euclidean geometry corresponded to zero curvature, *flat* space. These geometries could be characterized by their *metric,* the formula for the distance between two points.

These ideas may have led Gauss to a more general study of curved surfaces. He developed a beautiful formula for the amount of curvature, and proved that it gave the same result in any coordinate system. In this formulation, the curvature did not have to be constant: it could vary from one place to another.

In a move that is not unusual in mathematics, Gauss in middle age turned to practical applications. He assisted several surveying projects, the biggest being the triangulation of the region of Hanover. He did a lot of fieldwork, followed by data analysis. To aid the work he invented the heliotrope, a device for sending signals by reflected light. But when his heart began to show signs of failure, he stopped surveying and decided to spend his remaining years in Göttingen.

During this unhappy period a young Norwegian named Abel wrote to him about the impossibility of solving the quintic equation by radicals, but received no reply. Probably Gauss was too depressed even to look at the paper.

Around 1833, he became interested in magnetism and electricity, collaborating with the physicist Wilhelm Weber on a book, *General Theory of Terrestrial Magnetism,* published in 1839. They also invented a telegraph, linking Gauss's observatory to the physics laboratory where Weber worked, but the wires kept breaking, and other inventors came up with more practical designs. Then Weber was fired from Göttingen along with six others because they refused to swear allegiance to the new king of Hanover, Ernst August. Gauss was very upset by this, but his political conservatism and reluctance to make waves prevented him from raising any public protest, though he may have made efforts behind the scenes on Weber's behalf.

In 1845, Gauss produced a report on the pension fund for widows of Göttingen professors, examining the likely effect of a sudden increase in the number of members. He invested in railway and government bonds and amassed a tidy fortune.

After 1850, troubled by the onset of heart problems, Gauss cut back on work. The most important event of that period, for our story, was the habilitation thesis of his student Georg Bernhard Riemann. (In the German academic system, habilitation is the next step up after a PhD.) Riemann generalized Gauss's work on surfaces to multidimensional spaces, which he called "manifolds." In particular, he extended the concept of a metric, and found a formula for the curvature of a manifold. In effect, he created a theory of curved multidimensional spaces. Later this idea was to prove crucial in Einstein's work on gravity.

Gauss, now being regularly seen by his doctor, attended Riemann's public lecture on the topic and was impressed. As his health deteriorated further, he spent more and more time in bed but continued to write letters, read, and manage his investments. Early in 1855, Gauss died peacefully in his sleep, the greatest mathematical mind the world has ever known.

6

THE FRUSTRATED DOCTOR
AND THE SICKLY GENIUS

The first significant advance over Cardano's *The Great Art* came about halfway through the eighteenth century. Although the Renaissance mathematicians could solve cubics and quartics, their methods were basically a series of tricks. Each trick worked, but more, it seemed, by a series of coincidences than for any systematic reason. That reason was finally pinned down around 1770 by two mathematicians: Joseph-Louis Lagrange, a native of Italy who always considered himself French, and Alexandre-Théophile Vandermonde, who definitely was.

Vandermonde was born in Paris in 1735. His father wanted him to become a musician, and Vandermonde became proficient on the violin and followed a musical career. But in 1770 he became interested in mathematics. His first mathematical publication was about symmetric functions of the roots of a polynomial—algebraic formulas like the sum of all the roots, which do not change if the roots are interchanged. Its most original contribution was to prove that the equation $x^n - 1 = 0$, associated with the regular n-gon, can be solved by radicals if n is 10 or smaller. (Actually, it is solvable by radicals for any n.) The great French analyst Augustin-Louis Cauchy later cited Vandermonde as the first to realize that symmetric functions can be applied to the solution of equations by radicals.

In Lagrange's hands, this idea would form the starting point for an attack on all algebraic equations.

Lagrange was born in the Italian city of Turin and baptized Giuseppe Lodovico Lagrangia. His family had strong French links—his great-grandfather had been a captain in the French cavalry before moving to Italy to serve the duke of Savoy. When Giuseppe was quite young he started using Lagrange as a surname, but combined it with Lodovico or Luigi as his first name. His father was the treasurer of the Office of Public Works and Fortifications in Turin; his mother, Teresa Grosso, was a doctor's daughter. Lagrange was their first child out of an eventual total of eleven, but only two survived beyond childhood.

Although the family was in the upper levels of Italian society, they were strapped for cash, thanks to some bad investments. They decided that Lagrange should study law, and he attended the College of Turin. He enjoyed law and classics but found the mathematics classes, which consisted largely of Euclidean geometry, rather boring. Then he came across a book on algebraic methods in optics by the English astronomer Edmond Halley, and his opinion of mathematics changed dramatically. Lagrange was set on the course that would dominate his early research: the application of mathematics to mechanics, especially celestial mechanics.

He married a cousin, Vittoria Conti. "My wife, who is one of my cousins and who even lived for a long time with my family, is a very good housewife and has no pretensions at all," he wrote to his friend Jean le Rond D'Alembert, also a mathematician. He also confided that he did not want any children, an ambition he achieved.

Lagrange took a position in Berlin, wrote numerous research papers, and won the French Academy's annual prize on several occasions—sharing the 1772 prize with Euler, winning the 1774 prize for work on the dynamics of the Moon and the 1780 prize for work on the influence of planets on cometary orbits. Another of his loves was number theory, and in 1770 he proved a classic of the genre, the Four Squares Theorem, which asserts that every positive whole number is a sum of four perfect squares. For instance, $7 = 2^2 + 1^2 + 1^2 + 1^2$, $8 = 2^2 + 2^2 + 0^2 + 0^2$, and so on.

He became a member of the French Academy of Sciences and moved to Paris, where he remained for the rest of his life. He believed that it was wise to obey the laws of the country where you lived even if you disagreed with them, a point of view that probably helped him escape the fate of many other intellectuals during the French Revolution. In 1788 Lagrange published his masterpiece, *Analytical Mechanics,* which rewrote mechanics as a branch of analysis. He was proud that his massive book contained no diagrams whatsoever; in his view this made the logic more rigorous.

He married his second wife, Renée-Françoise-Adélaide Le Monnier, the daughter of an astronomer, in 1792. In August 1793, during the Reign of Terror, the Academy was shut down, and the only part that remained active was the commission on weights and measures. Many leading scientists were removed—the chemist Antoine Lavoisier, the physicist Charles Augustin Coulomb, and Pierre Simon Laplace. Lagrange became the new chairman of weights and measures.

At that point his Italian birth became a problem. The revolutionary government passed a law requiring any foreigner born in an enemy nation to be arrested. Lavoisier, who at that point retained some influence, arranged for Lagrange to be exempted from the new law. Soon afterward, a revolutionary tribunal condemned Lavoisier to death; he was guillotined the next day. Lagrange remarked, "It took only a moment to cause this head to fall, and a hundred years will not suffice to produce its like."

Under Napoleon, Lagrange was granted several honors: the Legion of Honor and Count of the Empire in 1808, and the Grand Cross of the Imperial Order of the Reunion in 1813. A week after receiving the Grand Cross, he was dead.

In 1770, the same year that he discovered his Four Squares Theorem, Lagrange embarked upon a vast treatise on the theory of equations, saying, "I propose in this memoir to examine the various methods found so far for the algebraic solution of equations, to reduce them to general principles, and to explain a priori why these methods succeed for the third and fourth degree, and fail for higher degrees." As Jean-Pierre Tignol put it in his book *Galois's Theory of Algebraic Equations,* Lagrange's "explicit aim is to determine not only *how* these methods work, but *why.*"

Lagrange reached a much deeper understanding of Renaissance methods than the methods' inventors had; he even proved that the general scheme that he had found to explain their success could not be extended to the fifth degree or higher. Yet he failed to take the further step of wondering whether any solution was *possible* in those cases. Instead, he tells us that his results "will be useful to those who will want to deal with the solution of the higher degrees, by providing them with various views to this end and above all by sparing them a large number of useless steps and attempts."

Lagrange had noticed that all of the special tricks employed by Cardano, Tartaglia, and others were based on one technique. Instead of trying to find the roots of the given equations directly, the idea was to transform the problem into the solution of some auxiliary equation whose roots were related to the original ones, but different.

The auxiliary equation for a cubic was simpler—a quadratic. This "resolvent quadratic" could be solved by the Babylonian method; then the solution of the cubic could be reconstructed by taking a cube root. This is exactly the structure of Cardano's formula. For a quartic, the auxiliary equation was also simpler—a cubic. This "resolvent cubic" could be solved by Cardano's method; then the solution of the quartic could be reconstructed by taking a fourth root—that is, a repeated square root. This is exactly the structure of Ferrari's formula.

We can imagine Lagrange's growing excitement. If the pattern continued, then the quintic equation would have a "resolvent quartic": solve that by Ferrari's method and then take a fifth root. And the process would continue, with the sextic having a resolvent quintic, solvable by what would be known as Lagrange's method. He would be able to solve equations of *any* degree.

Harsh reality brought him down to earth. The resolvent equation for the quintic was not a quartic but an equation of *higher* degree, a sextic. The method that had simplified the cubic and quartic equations *complicated* the quintic.

Mathematics does not progress by replacing difficult problems by even harder ones. Lagrange's unified method failed on the quintic. Still, he had not proved the quintic to be unsolvable, because there might exist different methods.

Why not?

To Lagrange, this was a rhetorical question. But one of his successors took it seriously, and answered it.

His name was Paolo Ruffini, and when I say that he "answered" Lagrange's rhetorical question, I am cheating slightly. He *thought* he had answered it, and his contemporaries never found anything wrong with his answer—partly because they never took his work seriously enough to really try. Ruffini spent his life believing that he had proved the quintic unsolvable by radicals. Only after his death did it turn out that his proof had a significant gap. It was easily overlooked among his pages and pages of intricate calculations; it was an "obvious" assumption, one that he had never even noticed he was making.

As every professional mathematician knows from bitter experience, it is very difficult to notice that you are making an unstated assumption, precisely because it is unstated.

Ruffini was born in 1765, the son of a doctor. In 1783 he enrolled at the University of Modena, studying medicine, philosophy, literature, and mathematics. He learned geometry from Luigi Fantini and calculus from Paolo Cassiani. When Cassiani moved on to a post with the Este family, managing their vast estates, Ruffini, though still a student, took charge of Cassiani's analysis course. He obtained a degree in philosophy, medicine, and surgery in 1788, adding a mathematics degree in 1789. Soon afterward, he took over a professorship from Fantini, whose eyesight was failing.

Events interfered with his academic work. Napoleon Bonaparte defeated the armies of Austria and Sardinia in 1796, turned his sights towards Turin, and captured Milan. Soon he had occupied Modena, and Ruffini was forced to become involved in politics. He had planned to go back to the university in 1798, but refused, on religious grounds, to swear allegiance to the republic. The resulting lack of employment left him more time to carry out his researches, and he focused on the vexed question of the quintic.

Ruffini convinced himself that there was a good reason why no one had managed to find a solution: there wasn't one. Specifically, there was no formula involving nothing more esoteric than radicals that would solve the general quintic. In his two-volume tome *General Theory of Equations,* published in 1799, he claimed to be able to prove this, asserting, "The algebraic solution of general equations of degree greater than four is always impossible. Behold a very important theorem which I believe I am able to assert (if I do not err): to present the proof of it is the main reason for publishing this volume. The immortal Lagrange, with his sublime reflections, has provided the basis of my proof."

The proof occupied more than 500 pages of largely unfamiliar mathematics. Other mathematicians found it somewhat daunting. Even today, no one is keen to wade through a very long and technical proof unless there is very good reason to do so. If Ruffini had announced a *solution* to the quintic, his peers would surely have made the effort. But you can understand their reluctance to devote hundreds of hours to the claim of a negative result.

Especially when it might be wrong. Few things are more annoying than finding an error on page 499 of a 500-page mathematics book.

Ruffini sent Lagrange a copy in 1801, and after a few months' silence he sent another copy, with a note: "If I have erred in any proof, or if I have said something which I believed new, and which is in reality not new, finally if I have written a useless book, I pray you point it out to me sincerely." Still no reply. He tried again in 1802. Nothing.

Several years passed without the recognition Ruffini felt was his due. Instead, vague rumors circulated, hinting that there were mistakes in his "proof," but since no one said what those mistakes might be, Ruffini was unable to defend himself. Eventually, he decided, no doubt correctly, that his proof was too complicated, and set about finding something simpler. He achieved this in 1803, writing, "In the present memoir, I shall try to prove the same proposition with, I hope, less abstruse reasoning and with complete rigor." The new proof fared no better. The world wasn't ready for Ruffini's insights or for the further proofs he published in 1808 and 1813. He never stopped trying to get his work recognized by the mathematical community. When Jean Delambre, who predicted the position of the planet Uranus, wrote a report on the state of mathematics since 1789, he included the sentence, "Ruffini proposes to prove that solving the quintic is impossible." Ruffini promptly replied, "I not only proposed to prove, but in reality did prove."

To be fair, a few mathematicians were happy with Ruffini's proof. Among them was Cauchy, who had a pretty poor track record when it came to giving credit where it was due, unless it was due to himself. In 1821, he wrote to Ruffini, "Your memoir on the general resolution of equations is a work which has always seemed to me worthy of the attention of mathematicians and which, in my judgment, proves completely the impossibility of solving algebraically equations of higher than the fourth degree." But by then the praise was far too late.

Around 1800 Ruffini started teaching applied mathematics in the city's military school. He continued to practice medicine, looking after patients from the poorest to the richest in society. In 1814, after the fall of Napoleon, he became rector of the University of Modena. The political situation was still extremely complex, and despite his personal skills, the great respect in which he was held, and his reputation for honesty, his time as rector must have been very difficult.

Simultaneously, Ruffini held the chairs of applied mathematics, practical medicine, and clinical medicine in the University of Modena. In 1817, there was a typhus epidemic and Ruffini continued to treat his patients until he caught the disease himself. He survived but never fully regained his health, and in 1819 he gave up his chair of clinical medicine. But he never gave up his scientific work, and in 1820 he published a scientific article on typhus based on his own experience as both physician and patient. He died in 1822, barely a year after Cauchy had written to praise his work on the quintic.

One reason Ruffini's work was not well received may have been its novelty. Like Lagrange, he based his investigations on the concept of a "permutation." A permutation is a way to rearrange some ordered list. The most familiar example is shuffling a pack of cards. The usual aim here is to achieve some random—that is, unpredictable—order. The number of different permutations of a pack of cards is huge, so the chance of predicting the outcome of random shuffling is negligible.

Permutations arise in the theory of equations because the roots of a given polynomial can be considered as a list. Some very basic features of equations are directly related to the effect of shuffling that list. The intuition is that the equation "does not know" the order in which you listed its roots, so permuting the roots should not make any important difference. In particular, the coefficients of the equation should be fully symmetric expressions in the roots—expressions that do not change when the roots are permuted.

But as Lagrange had appreciated, some expressions in the roots may be symmetric with respect to some permutations, but not others. These "partially symmetric" expressions are intimately associated with any formula for solving the equation. This feature of permutations was familiar to Ruffini's peers. Much less familiar was Ruffini's systematic use of another of Lagrange's ideas: that you can "multiply" two permutations to get another one by performing them in turn.

Consider the three symbols *a, b, c.* There are six permutations: *abc, acb, bac, bca, cab,* and *cba.* Take one of them, say *cba.* At first sight, this is just an ordered list formed from the three symbols. But we can also think of it as a *rule* for rearranging the original list *abc*. In this case, the rule is "reverse the order." And we can apply this rule not just to that list but to any list. Apply it to *bca,* say, and you get *acb.* So there is a sense in which *cba* × *bca* = *acb.*

This idea, which is central to our story, probably makes more sense if we draw some diagrams. Here are two diagrams for the permutations that rearrange *abc* into *cba* and *bca:*

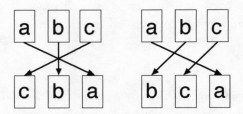

Two permutations of the symbols *a, b, c.*

We can combine the two rearrangements into one, by stacking these pictures on top of each other. There are two ways to do this:

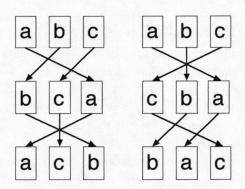

Multiplying permutations. The result depends on which comes first.

Now we can read off the result of "multiplying" the two permutations by writing down the bottom row, which here (left-hand picture) is *acb.* With this definition of "multiplication" (which is not the usual concept for multiplying numbers) we can make sense of the statement *cba* × *bca* = *acb.* The convention is that the first permutation in the product goes on the *bottom* of the stack. It matters, because we get a different answer if we swap the two layers of the stack. The right-hand picture shows that when the permutations are multiplied in the opposite order, the result is *bca* × *cba* = *bac*.

The essence of Ruffini's impossibility proof was to develop conditions that must be satisfied by any quintic whose roots can be expressed by radicals. If the general quintic does *not* satisfy these conditions, then it does not have that kind of root—and thus can't be solved by any natural extension of the methods that worked for the cubic and quartic.

Taking a leaf out of Lagrange's book, Ruffini homed in on symmetric functions of the roots and their relation to permutations. The quintic has five roots, and there are 120 permutations of five symbols. Ruffini realized that this system of permutations would have to possess certain structural features, inherited from any hypothetical formula for solutions of the quintic. If those features were absent, there could be no such formula. It was a bit like hunting a tiger in a muddy jungle. If there really was a tiger present, it would leave clear paw-prints in the mud. No paw-prints, no tiger.

By exploiting mathematical regularities of this new form of multiplication, Ruffini was able to prove—to his own satisfaction, at least—that the multiplicative structure of the 120 permutations is inconsistent with the symmetric functions that have to exist if the equation can be solved by radicals. And he did achieve something significant. Before Ruffini started working on the quintic, virtually every mathematician in the world was convinced that this equation could be solved; the only question was how. One exception was Gauss, who dropped hints that he thought no solution existed—but he also remarked that it wasn't a very interesting question, one of the few times his instincts let him down.

After Ruffini there seems to have been a general feeling that the quintic is not solvable by radicals. Very few thought Ruffini had *proved* this—but his work certainly made a lot of people feel rather doubtful that radicals were up to the job. This change of perception had an unfortunate side effect: mathematicians became much less interested in the whole issue.

Ironically, it later emerged that Ruffini's work had a major gap, but no one spotted it at the time. The skepticism of his contemporaries turned out to be justified, in a way. But the real breakthrough was the method:

Ruffini found the correct strategy; he just didn't use quite the right tactics. The subject needed a strategist who could also pay scrupulous attention to the minutiae of tactics. Now it got one.

After years of carrying out the good Lord's work, without complaint, as a pastor in some of the poorest and most remote regions of the Norwegian mountains, in 1784 Hans Mathias Abel received his just reward. He got himself appointed to the parish of Gjerstad, near the southern coast of Norway, not far from Oslo Fjord. Gjerstad wasn't exactly wealthy, but it was much richer than the places where he had previously ministered. His family finances would improve dramatically.

Spiritually, Pastor Abel's task was the same as ever: to look after his flock and do his best to keep them happy and virtuous. He came from a well-to-do family. His Danish great-grandfather had been a merchant who did a lucrative trade supplying the Norwegian army. His father, also a merchant, had been an alderman in the town of Bergen. Hans was proud but modest, not particularly intelligent but far from stupid, and prepared to speak his mind whatever the cost.

To help feed the poor of the parish, he grew new types of plant on his farm: flax, for making linen, and above all a new type of root vegetable, the ground apple, otherwise known as the potato. He wrote poetry, pottered about collecting information for a history of the area, and lived in harmony with his wife Elisabeth. His house was famous for the quality of its food; alcohol was never served. Drinking was a major social problem in Norway, and the pastor was determined to set his flock an example— though on one occasion he arrived in church as drunk as a newt, to show his parishioners how demeaning drunkenness was. He had two children, an unusually small family for the times: a daughter Margaretha, and a son Søren.

Margaretha was unexceptional, never married, and lived most of her life with her parents. Søren was altogether different: quick, intelligent, and original, with a taste for high society. He lacked his father's composure and sense of duty, and suffered for it. Still, he followed his father's profession, becoming first curate, then pastor; married Anne Marie Simonsen, the daughter of a family friend; and accepted a post in Finnøy, on the southwest coast. "The people round here are superstitious, but are filled with knowledge of the Bible," he wrote. "They support every erroneous opinion by misunderstood divine authority." Nevertheless, he enjoyed the job.

In 1801, Søren wrote to a friend, "My domestic joy has recently been increased, for on the third day of Christmas my wife presented me with a healthy son." This was Hans Mathias. A brother, Niels Henrik, arrived in the summer of 1802. From day one Niels suffered from ill health, and his mother had to spend a lot of time looking after him.

Military tensions were running high in Europe, and the combined state of Norway-Denmark was sandwiched between the major powers of England and France. Napoleon wanted to ally it with his cause, so when Britain came to an agreement with Sweden, Norway-Denmark instantly became an enemy of the British, who invaded. After three days, Norway-Denmark surrendered to save Copenhagen from destruction. Later, when Napoleon's grip on power was fading, his aide Jean Baptiste Bernadotte became king of Sweden. When Norway was ceded to Sweden, the Norwegian parliament, the Storting, was forced to accept Bernadotte as monarch.

The two boys were sent to the Cathedral School in Oslo in 1815. The mathematics teacher, Peter Bader, was the sort who motivated his students with serious physical violence. Nevertheless, both boys did well. Then in 1818, Bader gave one of the pupils—the son of a representative to the Storting—such a beating that the boy died. Amazingly, Bader was not tried, but he was replaced as mathematics teacher by Bernt Holmboe, who had been assistant to Christoffer Hansteen, the applied mathematics professor. This marked a turning point in Niels's mathematical career, because Holmboe allowed his pupils to tackle interesting problems outside the usual syllabus. Niels was permitted to borrow classic textbooks, among them some by Euler. "From now on," Holmboe later wrote, "[Niels] Abel devoted himself to mathematics with the most fervent eagerness and progressed in his science with the speed characteristic of a genius."

Shortly before finishing at school, Niels convinced himself that he had solved the quintic equation. Neither Holmboe nor Hansteen could find a mistake, so they transmitted the calculations to Ferdinand Degen, a prominent Danish mathematician, for possible publication by the Danish Academy of Science. Degen, too, found no errors in the work, but being an experienced hand who knew a trick or two, he asked Niels to try out the calculations on some specific examples. Niels quickly realized that something was amiss; he was disappointed, but relieved that he had not been allowed to make a fool of himself by publishing an erroneous result.

Søren's ambition and lack of tact now combined with embarrassing results. He read out a statement accusing two Storting representatives of unjustly imprisoning the manager of an ironworks, owned by one of them. This attack on their integrity created an uproar. It then transpired that the man concerned was unreliable, but Søren refused to apologize. Depressed and unhappy, he drank himself to death. At the funeral, Søren's widow, Anne Marie, became extremely drunk and took her favorite servant to bed. Next morning, she received several visiting officials—still in bed, with her lover beside her. An aunt wrote, "The poor boys, I feel sorry for them."

Niels graduated from the Cathedral School in 1821 and took the entrance examination to the University of Christiania (now Oslo). He received the highest possible grade in arithmetic and geometry and good grades in the rest of mathematics, and did terribly in everything else. Now desperately poor, he applied for a grant that would give him free accommodation and wood for the fire. He also sought a grant for living expenses, and some of the professors, recognizing his unusual talent, gave money to create a fellowship for him. Thus provided for, Niels devoted himself to mathematics and to solving the quintic, determined to make good his previous abortive attempt.

In 1823 Niels worked on elliptic integrals, an area of analysis that would be his most lasting monument, outclassing even his work on the quintic. He tried to prove Fermat's Last Theorem but found neither proof nor disproof, though he did show that any example that disproves the theorem must involve gigantic numbers.

In the summer of that year, he went to a ball, met a young woman, and asked her to dance. After several failed attempts, they both burst out laughing—neither of them had the foggiest idea how to dance. The lady was Christine Kemp, universally known as "Crelly," the daughter of a war commissar. Like Niels she had no money, and earned a living as a private tutor in everything from needlework to science. "She is not beautiful, has red hair and freckles, but she is a wonderful girl," he wrote. They fell in love.

These events gave Niels's mathematics a boost. Toward the end of 1823, he proved the quintic's impossibility—and unlike Ruffini's near miss, his did not have any gaps. Its strategy was similar to Ruffini's but with better tactics. Initially, Niels didn't know about Ruffini's work. Later, he certainly must have known of it because he alludes to its incompleteness. But even Niels did not put his finger on the precise gap in Ruffini's proof—even though his method turned out to be just what was needed to bridge the gap.

Niels and Crelly became engaged. To marry his sweetheart, Niels had to get a job—which meant his talents had to be recognized by Europe's leading mathematicians. Publishing his theory would not be enough: he had to beard the lions in their dens. And to do that, he needed enough money to travel.

After much effort, the University of Christiania was persuaded to grant Niels enough money for a research visit to Paris, where he would meet some of the world's leading mathematicians. In preparation for the trip, he decided that he needed printed copies of his best work. He believed that his impossibility proof for the quintic would impress his French peers; unfortunately, all of his work had been printed in Norwegian, in an obscure journal. He therefore decided that he should get his work on the theory of equations printed privately in French. Its title was "Memoir on algebraic equations, wherein one proves the impossibility of solving the general equation of the fifth degree."

To save on printing costs, Niels distilled his ideas down to their essentials, and the printed version ran a mere six pages. This was a lot less than Ruffini's 500 pages, but there are occasions in mathematics where brevity can make ideas more obscure. Many of the logical details—which in this area were crucial—had to be left out. The paper was a sketch, not a proof.

Niels introduced it by writing, "Mathematicians have generally been occupied with the problem of finding a general method for solving algebraic equations, and several have made attempts to prove the impossibility of it. I dare to hope, therefore, that mathematicians will receive favorably this article which has for its purpose to fill this lacuna in the theory of equations." It was a faint hope. Though he succeeded in visiting some mathematicians in Paris and getting them to agree to look at his paper, its reasoning was so compressed that most of them probably found it incomprehensible. Gauss filed his copy but never read it—when it was found after his death, the pages were still uncut.

Later, perhaps realizing his mistake, Abel produced two longer versions of his proof, giving more of the details. Having by this time heard of Ruffini, he wrote in these versions, "The first to attempt a proof of the impossibility of an algebraic solution of the general equation was the mathematician Ruffini; but his memoir is so complicated that it is difficult to judge the correctness of the argument. It appears to me that his reasoning is not always satisfactory." But like everyone else, he didn't say *why*.

Ruffini and Abel wrote their arguments in the formal mathematical language of the time, which was not well suited to the style of thinking required. Mathematics then was mainly concerned with specific, concrete ideas, whereas the key to the theory of equations is to think in rather general terms—about structures and processes rather than specific *things*. Thus their ideas were difficult for their contemporaries to grasp for reasons that went beyond language. But even for modern mathematicians, using the terminology of the period would make comprehension difficult.

Fortunately, we can grasp the essential features of their analysis by employing an architectural metaphor. One way to think about Ruffini's almost-proof, and of Abel's complete proof, is to imagine building a *tower*.

This tower has a single room on each floor, with a ladder that connects it to the room above. Each room contains a large sack. If you open the sack, millions of algebraic formulas spill out across the floor. At first sight, these formulas have no special structure and appear to have been harvested at random from the pages of algebra texts. Some are short, some long; some are simple, some extraordinarily complicated. A closer look, however, reveals family resemblances. The formulas in a given sack have lots of common features. The formulas in the sack in the room above have different common features. The higher we climb the tower, the more complicated the formulas in the sacks become.

The sack on the first floor, at ground level, contains all of the formulas that you can build by taking the coefficients of the equation and then adding them together, subtracting, multiplying, and dividing them—over and over, as many times as you like. In the world of algebraic formulas, once you have the coefficients, all of these "harmless" combinations come along pretty much free of charge.

To climb the ladder to the floor above, you must take some formula out of the sack, and use it to form a *radical*. It might be a square root, a cube root, a fifth root, whatever. But the formula whose root you are taking must have come from that sack. You can always take it to be a pth root where p is prime, because more complex roots can be built from prime ones, and this simple observation is surprisingly helpful.

Whichever root you decide to take, when you arrive on the second floor, you find a second sack, whose contents are initially identical to those of the sack of the first floor. But you open the sack, and throw in your new radical.

Formulas *breed*. When Noah landed his ark on Mount Ararat, he told all the creatures inside it to go forth and multiply. The formulas in the sack do more than that: they go forth and multiply, add, subtract, and divide. After a few seconds of frenzied activity, the sack on the second floor is bulging with all possible "harmless" combinations of the coefficients of the equation *and* your new radical. Compared to the sack on the first floor, there are many new formulas—but they all resemble each other; each of them includes your radical as a new component.

You do much the same to get to the third floor. Again you pick some formula from the new sack—just one—and form a new radical by taking some (prime) root of that formula. You carry your new radical up the ladder to the third floor, toss it in the sack, and wait for the formulas to carry out their mating rituals.

And so on. Each new floor introduces a new radical, and new formulas appear in the sack. At any stage, all of those formulas are built from the coefficients, together with any of the radicals introduced so far.

Eventually you reach the top floor of the tower. And you complete your quest—to solve the original equation by radicals—provided that, tucked away inside the sack in the attic, you can find at least one root of that equation.

There are many conceivable towers. They depend on which formulas you choose, and which radicals you take. Most fail dismally, and no hint of the desired root can be found. But if the quest is possible, if some formula built from successive radicals yields a solution, then the corresponding tower does indeed have a root in its attic. For the formula tells us exactly how to obtain that root by adjoining successive radicals. That is, it tells us exactly how to build the tower.

We can reinterpret the classic solutions of the cubic, the quartic, and even the Babylonian solution of quadratics in terms of these towers. We begin with the cubic, because this is complicated enough to be typical, but simple enough to be comprehensible.

Cardano Tower has only three floors.

The sack on the first floor contains the coefficients and all of their combinations.

The ladder to the second floor requires a square root. A very particular square root, that of a specific formula in the first sack. The sack on the second floor contains all combinations of this square root, together with the coefficients.

The ladder to the third floor, the attic, requires a cube root—again, a specific one. It is the cube root of a particular formula involving the coefficients and the square root that you used to reach the floor below. Does the sack in the attic contain a root of the cubic equation? It does, and the proof is Cardano's formula. The ascent of the tower is a success.

Ferrari Tower is taller; it has five floors.

The first floor, as always, has a sack that contains just the combinations formed by the coefficients. You reach the second floor by forming harmless combinations and then taking a suitable square root. You reach the

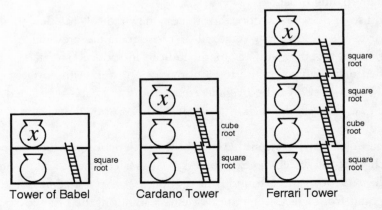

Solving the quadratic, cubic, and quartic.

third floor by forming harmless combinations and then taking a suitable cube root. You reach the fourth floor by forming harmless combinations and then taking a suitable square root. Finally, you clamber up to the fifth floor—the attic—by forming harmless combinations and then taking a suitable square root.

And now, the sack in the attic does indeed contain what you are seeking, a root of the quartic equation. Ferrari's formula provides the instructions for building precisely such a tower.

The Tower of Babel, which solves the quadratic, also fits the metaphor. But it turns out to be a stumpy tower with only two floors. The sack on the first floor contains just the combinations of the coefficients. A single carefully chosen square root conducts you to the floor above, the attic. Inside that sack is a root of the quadratic—in fact, both of them. The Babylonian procedure for solving quadratics, the formula you were taught at school, tells us so.

What about the quintic?

Suppose that a formula to solve the quintic by radicals really does exist. We don't know what it is, but we can infer a lot about it nonetheless. In particular, it must correspond to some tower. Let me call this hypothetical tower the Tower of Abel.

The Tower of Abel could contain hundreds of floors, and its ladders may involve all sorts of radicals—19th roots, 37th roots, we don't know. All we know for sure is that the sack on the first floor contains just the harmless combinations of coefficients. We fondly imagine that up in the attic, above the clouds, is a sack containing some root of the quintic.

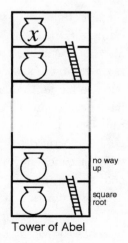

Tower of Abel

Why the quintic is unsolvable.

We ask how to climb the tower, and the mathematics tells us that there is only one way to get to the second floor. We have to take one particular square root. There is no other way up.

Well, not quite. We could take all sorts of other roots, build a huge, tall tower. But such a tower cannot have a root in its attic unless *some* floor corresponds to the particular square root that I am thinking about. And none of the previous floors will help you reach the attic; building them was a waste of time and money. So any sensible builder will go for that square root right at the start.

What do you need to climb the ladder to the third floor?

There *is* no ladder to the third floor. You can reach the second floor, but then you are stuck. And if you can't reach the third floor of the presumed tower, you certainly can't get to the attic and find a root in the sack.

In short, the Tower of Abel does not exist. All that exists is an abandoned attempt that peters out on the second floor; or perhaps a more elaborate structure with lots of unnecessary floors, which eventually peters out in exactly the same manner, for exactly the same reason. This is what Ruffini proved, save for one technical gap. Roughly speaking, he failed to prove that if harmless combinations of radicals live in the attic, then so do the radicals themselves.

Ruffini's proof and Abel's towers have clear similarities. But by using towers, Abel improved Ruffini's tactics and filled the gap he left. Between them, they proved that *no* radical tower climbs from the coefficients of the quintic to its roots. In architectural language, that tells us that there is no formula for the root of a quintic that uses nothing more elaborate than radicals. Solving the quintic by radicals is as impossible as climbing to the Moon by repeatedly standing on your own shoulders.

As Christmas 1828 drew near, Abel arranged to stay with his old friends Catharine and Niels Treschow in Froland. He was looking forward to visiting Crelly, who lived nearby. His doctor didn't think the trip was a good idea, because of the state of Abel's health. In a letter to Christoffer Hansteen's wife, Johanne, Catharine wrote, "If only you had been in town he might have been content to remain. But he tried to hide how ill he really was." In mid-December Abel headed for Froland, bundled up against the winter cold. He arrived on 19 December wearing every scrap of clothing he had with him, including socks over his arms and hands. Despite his coughs and cold shivers, he plowed ahead with his mathematics, happy to work in the Treschows' parlor surrounded by their children. He enjoyed the company.

Abel was still trying to land a permanent position. Even his temporary post at Oslo was in doubt. Over Christmas he focused his main efforts on securing the job in Berlin. His friend August Crelle, busy behind the scenes, had persuaded the Department of Education to create a mathematical institute and was angling for Abel to be appointed as one of its professors. He had obtained support from the scientific giant Alexander von Humboldt, together with a recommendation from Gauss and another from Adrien-Marie Legendre, a prominent member of the French Academy. Crelle advised the education minister that Abel was willing to accept a position in Berlin, but that the authorities should move quickly because he was in demand elsewhere, notably Copenhagen.

Abel was due to leave Froland for Oslo on 9 January, but his coughs and chills had worsened and he spent most of his time confined to his room. His intended in-laws, the Kemps, became very worried. On the morning of his planned departure he was coughing violently and spitting blood. The family doctor was immediately called to the house, and he prescribed bed rest and constant nursing. Crelly acted as nurse, and her loving attentions and various medications led to a distinct improvement. Within a few weeks Abel was allowed to sit in a chair for short periods. He had to be restrained from doing any mathematics.

Legendre wrote to say how impressed he was with Abel's work on elliptic functions, and urged the young man to publish his solution to the problem of deciding when an equation could be solved by radicals: "I urge you to let this new theory appear in print as quickly as you are able. It will be of great honor to you, and will universally be considered the greatest discovery which remained to be made in mathematics." While some prominent mathematicians, actively or through neglect, were hindering the publication of Abel's seminal works, his reputation in other quarters was growing fast.

Toward the end of February 1829, Abel's doctor realized that he was never going to recover, and the best he could hope for was to keep the illness at bay as long as possible. The doctor sent Abel's former teacher Bernt Holmboe a certificate, reporting the young man's state of health:

> . . . Shortly after his arrival at Froland Ironworks he suffered a severe attack of pneumonia with considerable expectoration of blood, which ceased after a brief period. But a chronic cough and great weakness have compelled him to rest in bed, where he must still remain; furthermore, he cannot be permitted to be exposed to the slightest variation in temperature.

More serious, the dry cough with stinging pains in the chest makes it very probable that he suffers from hidden chest and bronchial tubercles, which easily can result in a subsequent chest phthisis, partially on account of his constitution.

Due to this precarious state of health . . . it is most unlikely that he will be able to return to Oslo before the spring. Until then, he will be unable to discharge the duties of his office, even if the outcome of his illness should be the most desirable.

Crelle received the bad news in Berlin, and redoubled his efforts to secure Abel a position, advising the German minister that it would be good to transfer Abel to a warmer climate.

On 8 April, Crelle sent his protégé good news:

The Education Department has decided to call you to Berlin for an appointment . . . In what capacity you will be appointed and how much you will be paid I cannot tell you, for I do not know myself . . . I only wanted to hurry to let you hear the main news; you may be certain that you are in good hands. For your future you need no longer have any concern; you belong to us and are secure.

If only.

Abel was too ill to travel. He had to stay in Froland, where despite Crelly's nursing he became weaker and weaker, and his cough grew worse. He left his bed only to allow the sheets to be changed. When he tried to do some mathematics, he found he was unable to write. He began to dwell on the past, and his poverty, but he did not take his feelings out on the people he loved, remaining cooperative and good-natured to the very end.

Crelly naturally found it more and more difficult to hide her distress from her fiancé. Marie or Hanna kept her company at the bedside. Abel's worsening cough was stopping him from sleeping, and the family hired a nurse to look after him overnight so that Crelly could get some rest.

On the morning of 6 April, after a night of severe pain, Abel died. Hanna wrote, "He endured his worst agony during the night of 5 April. Toward morning he became more quiet, and in the forenoon, at 11 o'clock, he expired his last sigh. My sister and his fiancée were with him in the last moment, and saw his quiet passing into the arms of death."

Five days later, Crelly wrote to Catharine Hansteen's sister Henriette Fridrichsen, asking her to tell Catharine the sad news. "My dearest love, yes, only duty could make me demand this, for I owe your sister, Fru Hansteen, so much. I take the pen with trembling hand to ask you to inform her that she has lost a kind, devout son who loved her so infinitely.

"My Abel is dead! . . . I have lost all on earth. Nothing, nothing have I left. Pardon me, the unfortunate can write no more. Ask her to accept the enclosed lock of my Abel's hair. That you will prepare your sister for this in the most lenient way asks your miserable C. Kemp."

7

THE LUCKLESS REVOLUTIONARY

Mathematicians are never satisfied.

Whenever a problem is solved it only raises new questions. Soon after Abel's death, his proof that some quintics can't be solved by radicals started to become recognized. But Abel's work was just the start. Although all previous attempts to solve *all* quintics had ground to a halt, a few very clever mathematicians had proved that some quintics *can* be solved by radicals. Not just obvious ones, like $x^5 - 2 = 0$, where $x = \sqrt[5]{2}$, but surprising ones like $x^5 + 15x + 12 = 0$, though the solution is too complicated to state here.

This was a puzzle. If some quintics are solvable and some not, what distinguishes one kind from the other?

The answer to this question changed the course of mathematics, and mathematical physics. Even though the answer was given more than 170 years ago, it is still yielding important new discoveries. In retrospect, it is astonishing how far-reaching are the consequences of an innocent question about the internal structure of mathematics. Solving quintics, it appeared, had no practical use whatsoever. If some problem in engineering or astronomy involved a quintic, there were numerical methods to determine a solution to as many decimal places as were needed. The solvability—or not—of a quintic by radicals was a classic example of "pure" mathematics, of questions asked for reasons that interested no one but mathematicians.

How wrong you can be.

Abel had discovered an obstacle to the solution of certain quintics by radicals. He had proved that this obstacle genuinely prevented such solutions existing for at least some quintics. The next step forward, the pivot upon which our entire story revolves, was made by someone who looked the gift horse firmly in the teeth and asked the kind of question that mathematicians cannot resist when some major problem has been solved. "Yes, that's all very nice . . . but why does it *really* work?"

The attitude may seem rather negative, but time and time again it has proved its worth. The underlying philosophy is that most mathematical problems are too difficult for anyone to solve. So when somebody manages to solve something that has baffled all predecessors, merely celebrating the great solution is not enough. Either the solver got lucky (mathematicians do not believe in that sort of luck) or some special reason made the solution possible. And if it proves possible to understand the *reason* . . . why, lots of other problems might yield to similar methods.

So while Abel was polishing off the specific question, "Can every quintic be solved?" and getting a clear "no," an even deeper thinker was wrestling with a far more general issue: which equations can be solved by radicals, and which cannot? To be fair, Abel had begun to think along those lines, and might have found the answer if tuberculosis had spared him.

The person who was to change the course of mathematics and science was Évariste Galois, and his life story is one of the most dramatic, and also the most tragic, in the history of mathematics. His magnificent discoveries were very nearly lost altogether.

If Galois had not been born, or if his work had really been lost, someone would no doubt have made the same discoveries eventually. Many mathematicians had voyaged across the same intellectual territory, missing the great discovery by a whisker. In some alternative universe, someone with Galois's gifts and insights (perhaps a Niels Abel who avoided tuberculosis for a few more years) would eventually have penetrated the same circle of ideas. But in this universe, it was Galois.

He was born on 25 October 1811, in Bourg-la-Reine, in those days a small village on the outskirts of Paris. Now it is a suburb in the département Hau-de-Seine, at the intersection of the N20 and the D60 highways. The D60 is now named Avenue Galois. In 1792, the village of Bourg-la-Reine had been renamed Bourg-l'Égalité, a name that reflected the era's political turmoil and its ideology: "Queen Town" had given way to "Equality Town." In 1812, the name reverted to Bourg-la-Reine, but revolution was still in the air.

The father, Nicolas-Gabriel Galois, was a republican and leader of the village Liberal Party—Liberté in the town of Egalité—whose main policy was the abolition of the monarchy. When, in a fudged compromise of 1814, King Louis XVIII was returned to the throne, Nicolas-Gabriel became the town mayor, which cannot have been a comfortable office for someone of his political leanings.

The mother, Adelaide-Marie, was born to the Démante family. Her father was a jurisconsult, a paralegal expert whose job was to offer opinions about legal cases. Adelaide-Marie was a fluent reader of Latin and passed her classical education on to her son.

For his first twelve years, Évariste remained at home, educated by his mother. He was offered a place at the college of Reims when he was ten, but his mother seems to have thought it too early for him to leave home. But in October 1823, he started attending the Collège de Louis-le-Grand, a preparatory school. Soon after Évariste arrived, the students refused to chant in the school chapel, and the young Galois saw at first hand the fate of would-be revolutionaries: a hundred pupils were promptly expelled. Unfortunately for mathematics, the lesson did not deter him.

For his first two years he was awarded first prize in Latin, but then he became bored. In consequence, the school insisted that he repeat his classes to improve his performance, but of course this made him even more bored, and things went from bad to worse. What saved Galois from the slippery slope to oblivion was mathematics, a subject with enough intellectual content to retain his interest. And not just any mathematics: Galois went straight to the classics: Legendre's *Elements of Geometry.* It was a bit like a modern physics student starting out by reading the technical papers of Einstein. But in mathematics there is a kind of threshold effect, an intellectual tipping point. If a student can just get over the first few humps, negotiate the notational peculiarities of the subject, and grasp that the best way to make progress is to *understand* the ideas, not just learn them by rote, he or she can sail off merrily down the highway, heading for ever more abstruse and challenging ideas, while an only slightly duller student gets stuck at the geometry of isosceles triangles.

Just how hard Galois had to work to understand Legendre's seminal work is open to dispute, but in any case it did not daunt him. He started to read the technical papers of Lagrange and Abel; not surprisingly, his later work concentrated on their areas of interest, in particular the theory of equations. Equations were possibly the only things that really grabbed

Galois's attention. His ordinary schoolwork suffered in proportion to his devotion to the works of the mathematical greats.

At school, Galois was untidy, a habit he never lost. He baffled his teachers by solving problems in his head instead of "showing his work." This is a fetish of mathematics teachers that afflicts many a talented youngster today. Imagine what would happen to a budding young footballer if every time he scored a goal, the coach demanded that he write out the exact sequence of tactical steps he followed, or else the goal would be invalid. There was no such sequence. The player saw an opening and put the ball where anyone who understood the game would know it had to go.

So it is with able young mathematicians.

Ambition led Galois to aim high: he wanted to continue his studies at one of the most prestigious institutions in France, the École Polytechnique, the breeding ground of French mathematics. But he ignored the advice of his mathematics teacher, who tried to make the young man work in a systematic manner, show his work, and generally make it possible for the examiners to follow his reasoning. Fatally underprepared and overconfident, Évariste took the entrance examination—and failed.

Twenty years later, an influential French mathematician named Orly Terquem, who edited a prestigious journal, offered an explanation for Galois's failure: "A candidate of superior intelligence is lost with an examiner of inferior intelligence. Because they do not understand me, *I* am a barbarian." A modern commentator, more aware of the need for communication skills, would temper that criticism with the observation that a student of superior intelligence has to make allowances for those less able. Galois did not help his case by being uncompromising.

So Galois remained at Louis-le-Grand, where he had a rare piece of good fortune. A teacher named Louis-Paul Richard recognized the young man's talent, and Galois enrolled in an advanced mathematics course under Richard's tuition. Richard formed the opinion that Galois was so talented that he should be admitted to the École Polytechnique without being examined. Very likely, Richard had an idea of what would happen if Galois were to take the examination. There is no evidence that Richard ever explained his view to the École Polytechnique. If he did, they took no notice.

✳

By 1829, Galois had published his first research paper, a competent but pedestrian article on continued fractions. His unpublished work was more ambitious: he had been making fundamental contributions to the theory of equations. He wrote up some of his results and sent them to the French Academy of Sciences, for possible publication in their journal. Then, as now, any paper submitted for publication would be sent to a referee, an expert in the field concerned, who made recommendations about the novelty, value, and interest of the work. In this case the referee was Cauchy, then probably France's leading mathematician. Having already published in areas close to those involved in Galois's paper, he was a natural choice.

Unfortunately, he was also extremely busy. There is a prevalent myth that Cauchy lost the manuscript; some sources suggest that he threw it away in a fit of pique. The truth seems more prosaic. There is a letter from Cauchy to the Academy, dated 18 January 1830, in which he apologizes for not presenting a report on the work of "young Galoi," explains that he was "indisposed at home," and also mentions a memoir of his own.

This letter tells us several things. The first is that Cauchy had not thrown Galois's manuscript away but still had it six months after submission. The second is that Cauchy must have read the manuscript and decided that it was important enough to be worth drawing to the Academy's attention.

But when Cauchy turned up at the next meeting he presented only his own paper. What had happened to Galois's manuscript?

The French historian René Taton has argued that Cauchy was impressed by Galois's ideas—perhaps a little too impressed. So instead of reading the work to the Academy as originally intended, he advised Galois to write a more extensive and presumably much improved exposition of the theory, to be submitted for the Grand Prize in Mathematics, a major honor. There is no documentary evidence to confirm this claim, but we do know that in February 1830 Galois submitted just such a memoir for the Grand Prize.

We cannot know exactly what was in this document, but its general contents can be inferred from Galois's surviving writings. It is clear that history might have been very different if the far-reaching implications of his work had been fully appreciated. Instead, the manuscript just vanished.

One possible explanation appeared in 1831 in *The Globe,* a journal published by the Saint-Simonians, a neo-Christian socialist movement. *The*

Globe reported a court case in which Galois was accused of publicly threatening the life of the king, and suggested that "This memoir . . . deserved the prize, for it could resolve some difficulties that Lagrange had failed to do. Cauchy had conferred the highest praise on the author about this subject. And what happened? The memoir is lost and the prize is given without the participation of the young savant."

The big problem here is to decide the factual basis of the article. Cauchy had fled the country in September 1830 to avoid the revolutionaries' anti-intellectual attentions, so the article cannot have been based on anything he had said. Instead, it looks as though the source was Galois himself. Galois had a close friend, Auguste Chevalier, who had invited him to join a Saint-Simonian commune. It seems likely that Chevalier was the reporter—Galois was otherwise engaged at the time, on trial for his life—and if so, the story must have come from Galois. Either he made it all up, or Cauchy had indeed praised his work.

Let us return to 1829. On the mathematical front, Galois was becoming increasingly frustrated by the apparent inability of the mathematical community to give him the recognition he craved. Then his personal life began to fall to pieces.

All was not well in the village of Bourg-la-Reine. The village mayor, Galois's father, Nicolas, became involved in a nasty political dispute, which enraged the village priest. The priest took the decidedly uncharitable step of circulating malicious comments about Nicolas's relatives and forging Nicolas's own signature on them. In despair, Nicolas committed suicide by suffocating himself.

This tragedy happened just a few days before Galois's final opportunity to pass the entrance examination for the École Polytechnique. It did not go well. Some accounts have Galois throwing the blackboard eraser into the examiner's face—it was probably a cloth, not a lump of wood, but even so, the examiner would not have been favorably impressed. In 1899, J. Bertrand provided some details that suggest that Galois was asked a question he had not anticipated, and lost his temper.

For whatever reason, Galois failed the entrance exam, and now he was in a bind. Having been utterly confident that he would pass—he really does seem to have been an arrogant young man—he had not bothered to prepare for the exams to enter the only alternative, the École Préparatoire.

Nowadays, this institution, renamed the École Normale, is more prestigious than the Polytechnique, but in those days it came a poor second. Galois hastily boned up on the necessary material, passed his mathematics and physics with flying colors, made a mess of his literature exam, and was accepted anyway. He obtained qualifications in both science and letters at the end of 1829.

As I mentioned, in February 1830 Galois submitted a memoir on the theory of equations to the Academy for the Grand Prize. The secretary, Joseph Fourier, took it home to give it the once-over. The ill-fortune that constantly dogged Galois's career struck again: Fourier promptly died, leaving the memoir unread. Worse, the manuscript could not be found among his papers. However, there were three other committee members in charge of the prize: Legendre, Sylvestre-François Lacroix, and Louis Poinsot. Maybe one of them lost it.

Galois, not surprisingly, was furious. He became convinced that what had happened was a conspiracy of mediocre minds to stifle the efforts of genius; he quickly found a scapegoat, the oppressive Bourbon regime. And he wanted to play a role in its destruction.

Six years earlier, in 1824, King Charles X had come to the throne of France, following Louis XVIII, but he was far from popular. The liberal opposition did well in the 1827 elections and even better in 1830, gaining a majority. Charles, facing the imminent prospect of forced abdication, attempted a coup; on 25 July he issued a proclamation suspending freedom of the press. He misread the mood of the people, who promptly rose in revolt, and after three days a compromise was reached: Charles was replaced as king by the duke of Orléans, Louis-Philippe.

The students of the École Polytechnique, the university Galois had hoped to attend, played a significant role in these events, demonstrating on the streets of Paris. And where was the arch antimonarchist Galois during this fateful period? Locked away inside the École Préparatoire along with his fellow students. The Director, Guigniault, had decided to play safe.

Galois was so incensed at being denied his place in history that he wrote a blistering attack on Guigniault in the *Gazette des Écoles:*

The letter which M. Guigniault placed in the lycée yesterday, on the account of one of the articles in your journal, seemed to me most improper. I had thought that you would welcome eagerly any way of exposing this man.

Here are the facts, which can be vouched for by forty-six students.

On the morning of 28 July, when several students of the École Normale wanted to join in the struggle, M. Guigniault told them, twice, that he had the power to call the police to restore order in the school. The police on 28 July!

The same day, M. Guigniault told us with his usual pedantry: "There are many brave men fighting on both sides. If I were a soldier, I would not know which to decide. Which to sacrifice, liberty or LEGITIMACY?"

There is the man who next day covered his hat with an enormous tricolor cockade [a symbol of the republicans]. There are our liberal doctrines!

The editor published the letter but removed the author's name from it. The director promptly expelled Galois for publishing an anonymous letter.

Galois retaliated by joining the Artillery of the National Guard, a paramilitary organization that was a hotbed of republicanism. On 21 December 1830, this unit, very probably including Galois, was stationed in the vicinity of the Louvre. Four ex-ministers had gone on trial, and the public mood was ugly: they wanted the men executed, and were prepared to riot if they were not. But just before the verdict was announced, the Artillery of the National Guard was withdrawn and replaced by the regular National Guard, together with other soldiers who were loyal to the King. The verdict of a jail sentence was announced, the riot failed to materialize, and ten days later, Louis-Philippe disbanded the Artillery of the National Guard as a security risk. Galois was having no more success as a revolutionary than he had had as a mathematician.

Practical issues now became more urgent than politics: he needed to make a living. Galois set himself up as a private mathematics tutor, and forty students signed up for a course of advanced algebra. We know that Galois was not a good written expositor, and it's reasonable to guess that his teaching was no better. Probably his classes were laced with political commentary; almost certainly they were too difficult for ordinary mortals. At any rate, the enrollment rapidly dwindled.

Galois had still not given up on his mathematical career, and he submitted yet a third version of his work to the Academy, entitled *On the Conditions of Solvability of Equations by Radicals*. With Cauchy having fled Paris, the referees were Siméon Poisson and Lacroix. When two months passed without any response, Galois wrote to ask what was happening. No one replied.

By the spring of 1831, Galois was behaving ever more erratically. On 18 April the mathematician Sophie Germain, who had greatly impressed Gauss when she first began her research in 1804, wrote a letter about Galois to Guillaume Libri: "They say he will go completely mad, and I fear this is true." Never the most stable person, he was now verging on full-blooded paranoia.

That month, the authorities arrested nineteen members of the Artillery because of the events at the Louvre and put them on trial for sedition, but the jury acquitted the men. The Artillery held a celebration on 9 May in which about two hundred Republicans assembled for a banquet at the restaurant Vendanges des Bourgogne. Every one of them wanted to see Louis-Philippe overthrown. The novelist Alexandre Dumas, who was present, wrote, "It would be difficult to find in all Paris, two hundred persons more hostile to the government than those to be found reunited at five o'clock in the afternoon in the long hall on the ground floor above the garden." As the event became more and more riotous, Galois was seen with a glass in one hand and a dagger in the other. The participants interpreted this gesture as a threat to the king, approved wholeheartedly, and ended up dancing in the streets.

The next morning, Galois was arrested at his mother's house—which suggests that there had been a police spy at the banquet—and charged with threatening the king's life. For once he seems to have learned some political sense, because at his trial he admitted everything, with one modification: he claimed that he had proposed a toast to Louis-Philippe, and had gestured with the dagger while adding the words, "if he turns traitor." He lamented that these vital words had been drowned in the uproar.

Galois made it clear, however, that he *did* expect Louis-Philippe to betray the people of France. When the prosecutor asked whether the accused could "believe this abandonment of legality on the part of the king," Galois responded, "He will soon turn traitor if he has not done so already." Pushed further, he left no doubt as to his meaning: "The trend in government can make one suppose that Louis-Philippe will betray one day if he hasn't already." Despite this, the jury acquitted him. Perhaps they felt as he did.

On 15 June, Galois was at liberty. Three weeks later, the Academy reported on his memoir. Poisson had found it "incomprehensible." The report itself said this:

> We have made every effort to understand Galois's proof. His reasoning is not sufficiently clear, not sufficiently developed, for us to judge its correctness, and we can give no idea of it in this report. The author announces that the proposition which is the special object of this memoir is part of a general theory susceptible of many applications. Perhaps it will transpire that the different parts of a theory are mutually clarifying, are easier to grasp together rather than in isolation. We would then suggest that the author should publish the whole of his work in order to form a definitive opinion. But in the state which the part he has submitted to the Academy now is, we cannot propose to give it approval.

The most unfortunate feature of this report is that it may well have been entirely fair. As the referees pointed out:

> [The memoir] does not contain, as [its] title promised, the condition of solvability of equations by radicals; indeed, assuming as true M. Galois's proposition, one could not derive from it any good way of deciding whether a given equation of prime degree is solvable or not by radicals, since one would first have to verify whether this equation is irreducible and next whether any of its roots can be expressed as a rational fraction of two others.

The final sentence here refers to a beautiful criterion for solvability by radicals of equations of prime degree that was the climax of Galois's memoir. It is indeed unclear how this test can be applied to any specific equation, because you need to know the roots before the test can be applied. But without a formula, in what sense can you "know" the roots? As Tignol says, "Galois's theory did not correspond to what was expected; it was too novel to be readily accepted." The referees wanted some kind of condition on the *coefficients* that determined solubility; Galois gave them a condition on the *roots*. The referees' expectation was unreasonable. No simple criterion based on the coefficients has ever been found, nor is one remotely likely. But hindsight cannot help Galois.

On 14 July, Bastille Day, Galois and his friend Ernest Duchâtelet were at the head of a Republican demonstration. Galois was wearing the uniform of the disbanded Artillery and carrying a knife, several pistols, and a loaded rifle. It was illegal to wear the uniform, and also to be armed. Both men were arrested on the Pont-Neuf, and Galois was charged with the lesser offense of illegally wearing a uniform. They were sent to the jail at Sainte-Pélagie to await trial.

While in jail, Duchâtelet drew a picture on the wall of his cell showing the king's head, labeled as such, lying next to a guillotine. This presumably did not help their cause.

Duchâtelet stood trial first; then it was Galois's turn. On 23 October he was tried and convicted; his appeal was turned down on 3 December. By this time he had spent more than four months in jail. Now he was sentenced to another six months. He worked for a while on his mathematics; then in the cholera epidemic of 1832 he was transferred to a hospital and later put on parole. Along with his freedom he experienced his first and only love affair, with a certain "Stéphanie D," as his doodles identify her.

From this point on it takes a lot of guesswork to interpret the scanty historical record. For a time, no one knew Stéphanie's surname or what sort of person she was. This mystery added to her romantic image. Galois wrote her full name on one of his manuscripts, but at some later point he scrawled all over it, rendering it illegible. Forensic work by the historian Carlos Infantozzi, who examined the manuscript very carefully, revealed the lady as Stéphanie-Felicie Poterin du Motel. Her father, Jean-Louis Auguste Poterin du Motel, was resident physician at the Sieur Faultrier, where Galois spent the last few months of his life.

We don't know what Jean-Louis thought of the relationship, but it seems unlikely that he approved of a penniless, unemployed, dangerously intense young man with extremist political views and a criminal record paying court to his daughter.

We do know a little about Stéphanie's opinions, but only through some scribbled sentences that Galois presumably copied from her letters. There is much mystery surrounding this interlude, which has a crucial bearing on subsequent events. Apparently, Galois was rejected and took it very badly, but the circumstances cannot be determined. Was it all in his mind—an infatuation that was never reciprocated? Did Stéphanie encourage his advances? Did she then get cold feet? The very characteristics likely to repel her father might have been distinctly attractive to the daughter.

As far as Galois was concerned, the relationship was certainly serious. In May, he wrote to his close friend Chevalier, "How can I console myself when in one month I have exhausted the greatest source of happiness a man can have?" On the back of one of his papers he made fragmentary copies of two letters from Stéphanie. One begins, "Please let us break up this affair," which suggests that there was something to break up. But it continues, "and do not think about those things which did not exist and which never would have existed," giving the contrary impression. The other contains the following sentences: "I have followed your advice and I have thought over what . . . has . . . happened . . . In any case, Sir, be assured there never would have been more. You are assuming wrongly and your regrets have no foundation."

Whether he imagined the whole thing and his feelings were never reciprocated, or he initially received some form of encouragement only to be subsequently rejected, it looks as though Galois suffered the worst kind of unrequited love. Or was the whole affair perhaps more sinister? Shortly after the breakup with Stéphanie, or what Galois interpreted as a breakup, someone challenged him to a duel. The ostensible reason was that this person objected to Galois's advances toward the young lady, but yet again the circumstances are veiled in mystery.

The standard story was one of political intrigue. Writers like Eric Temple Bell and Louis Kollros tell us that Galois's political opponents found his infatuation with Mlle. du Motel to be the perfect excuse to eliminate their enemy on a trumped-up "affair of honor." One rather wild suggestion is that Galois was the victim of a police spy.

These theories now seem implausible. Dumas states in his *Memoirs* that Galois was killed by Pescheux D'Herbinville, a fellow Republican whom Dumas described as "a charming young man who made silk-paper cartridges which he would tie up with silk ribbons." These were an early form of cracker, of the kind now familiar at Christmas. D'Herbinville was something of a hero to the peasantry, having been one of the nineteen Republicans acquitted on charges of conspiring to overthrow the government. Certainly he was not a spy for the police, because Marc Caussidière named all such spies in 1848 when he became chief of police.

The police report on the duel suggests that the other participant was one of Galois's revolutionary comrades, and the duel was exactly what it appeared to be. This theory is largely borne out by Galois's own words on the matter: "I beg patriots and my friends not to reproach me for dying otherwise than for my country. I die the victim of an infamous coquette. It is in a miserable brawl that my life is extinguished. Oh! why die for so trivial a thing, for something so despicable! . . . Pardon for those who have killed me, they are of good faith." Either he was unaware that he was the victim of a political plot, or there was no plot.

It does appear that Stéphanie was at least a proximate cause of the duel. Before departing for the engagement, Galois left some final doodles on his table. They include the words "Une femme," with the second word scribbled out. But the ultimate cause is as opaque as much else in this tale.

The mathematical story is much clearer. On 29 May, the eve of the duel, Galois wrote to Auguste Chevalier, outlining his discoveries. Chevalier eventually published the letter in the *Revue Encyclopédique*. It sketches the connection between groups and polynomial equations, stating a necessary and sufficient condition for an equation to be solvable by radicals.

Galois also mentioned his ideas about elliptic functions and the integration of algebraic functions, and other things too cryptic to be identifiable. The scrawled comment "I have no time" in the margins has given rise to another myth: that Galois spent the night before the duel frantically writing out his mathematical discoveries. But that phrase has next to it "(Author's note)," which hardly fits such a picture; moreover, the letter was an explanatory accompaniment to Galois's rejected third manuscript, complete with a marginal note added by Poisson.

The duel was with pistols. The postmortem report states that they were fired at 25 paces, but the truth may have been even nastier. An article from the 4 June 1832 issue of *Le Precursor* reported:

Paris, 1 June—A deplorable duel yesterday has deprived the exact sciences of a young man who gave the highest expectations, but whose celebrated precocity was lately overshadowed by his political activities. The young Évariste Galois . . . was fighting with one of his old friends, a young man like himself, like himself a member of the Society of Friends of the People, and who was known to have figured equally in a political trial. It is said that love was the cause of the combat. The pistol was the chosen weapon of the adversaries, but because of their old friendship they could not bear to look at one another and left the decision to blind fate. At point-blank range they were each armed with a pistol and fired. Only one pistol was charged. Galois was pierced through and through by a ball from his opponent; he was taken to the Hospital Cochin where he died in about two hours. His age was 22. L.D., his adversary, is a bit younger.

Could "L.D." refer to Pescheux d'Herbinville? Perhaps. The letter D is acceptable because of the variable spelling of the period; the L may have been a mistake. The article is unreliable on details: it gets the date of the duel wrong, and also the day Galois died and his age. So the initial might also be wrong.

The cosmologist and writer Tony Rothman has a more convincing theory. The person who best fits the description here is not d'Herbinville but Duchâtelet, who was arrested with Galois on the Pont-Neuf. Galois's biographers Robert Bourgne and Jean-Pierre Azra give Duchâtelet's Christian name as "Ernest," but that might be wrong, or again the L may be wrong. To quote Rothman, "We arrive at a very consistent and believable picture of two old friends falling in love with the same girl and deciding the outcome by a gruesome version of Russian roulette."

This theory is also consistent with a final horrific twist to the tale. Galois was hit in the stomach, a wound that was almost always fatal. If the duel was at point-blank range, this is no great surprise; if at 25 paces, it is the final example of his cursed luck.

He did not die two hours later, as *Le Precursor* says, but in the Hospital Cochin the next day, on 31 May. The cause of death was peritonitis, and he refused the office of a priest. On 2 June 1832 Galois was buried in the common ditch at the cemetery of Montparnasse.

His letter to Chevalier ended with these words: "Ask Jacobi or Gauss publicly to give their opinion, not as to the truth, but as to the importance of these theorems. Later there will be, I hope, some people who will find it to their advantage to decipher all this mess."

But what did Galois actually accomplish? What was the "mess" referred to in his final letter?

The answer is central to our tale, and not easily stated in a few sentences. Galois introduced a new point of view into mathematics, he changed its content, and he took a necessary but unfamiliar step into abstraction. In Galois's hands, mathematics ceased to be the study of numbers and shapes—arithmetic, geometry, and ideas that developed out of them like algebra and trigonometry. It became the study of *structure*. What had been a study of *things* became a study of *processes*.

We should not give Galois all the credit for this transformation. He was riding a wave that had been set in motion by Lagrange, Cauchy, Ruffini, and Abel. But he rode it with such skill that he made it his own; he was the first person seriously to appreciate that mathematical questions could sometimes be best understood by transporting them into a more abstract realm of thought.

It took a while for the beauty and value of Galois's results to percolate into the general mathematical consciousness. In fact they were very nearly lost. They were rescued by Joseph-Louis Liouville, the son of a captain in Napoleon's army who became a professor at the Collège de France. Liouville spoke to the French Academy—the body that had mislaid or rejected Galois's three memoirs—in the summer of 1843. "I hope to interest the Academy," he began, "in announcing that among the papers of Évariste Galois I have found a solution, as precise as it is profound, of this beautiful problem: whether or not there exists a solution by radicals . . ."

If Liouville had not bothered to wade through the luckless revolutionary's often untidy and confusing manuscripts, and had not devoted considerable time and effort to puzzling out what the author intended, the manuscripts might well have been thrown out with the rubbish, and group theory would have had to await some later rediscovery of the same ideas. So mathematics owes Liouville an enormous debt.

As understanding of Galois's methods grew, a new and powerful mathematical concept came into being: that of a group. An entire branch of mathematics, a calculus of symmetry called group theory, came into being and has since invaded every corner of mathematics.

Galois worked with groups of permutations—ways to rearrange a list of objects. In his case, the objects were the roots of an algebraic equation. The simplest interesting example is a general cubic equation, with three roots *a, b,* and *c*. Recall that there are six ways to permute these symbols, and that—following Lagrange and Ruffini—we can *multiply* any two permutations by performing them in turn. We saw, for example, that *cba* × *bca* = *acb*. Proceeding in this way, we can build up a "multiplication table" for all six permutations. It's easier to see what's going on if we assign names to each permutation, say by letting $I = abc$, $R = acb$, $Q = bac$, $V = bca$, $U = cab$, and $P = cba$. Then the multiplication table looks like this:

	I	**U**	**V**	**P**	**Q**	**R**
I	I	U	V	P	Q	R
U	U	V	I	R	P	Q
V	V	I	U	Q	R	P
P	P	Q	R	I	U	V
Q	Q	R	P	V	I	U
R	R	P	Q	U	V	I

Multiplication table for the six permutations of the roots of a cubic equation.

Here the entry in row X and column Y is the product XY, which means "do Y, then do X."

Galois realized that a very simple and obvious feature of this table is crucially important. The product of any two permutations is itself a permutation—the only symbols appearing in the table are I, U, V, P, Q, R. Some smaller collections of permutations have the same "group property": the product of any two permutations in the collection is also in the collection. Galois called such a collection of permutations a *group*.

For example, the collection $[I, U, V]$ gives a smaller table:

	I	**U**	**V**
I	I	U	V
U	U	V	I
V	V	I	U

Multiplication table for a subgroup of three permutations.

and only those three symbols appear. When, as here, one group is part of another, we call it a *subgroup*.

Other subgroups, namely [I, P], [I, Q], and [I, R], contain only two permutations. There is also the subgroup [I] that contains only I. It can be proved that the six subgroups just listed are the *only* subgroups of the group of all permutations on three symbols.

Now, said Galois (though not in this language), if we choose some cubic equation, we can look at its symmetries—those permutations that preserve all algebraic relations between the roots. Suppose, for example, that $a + b^2 = 5$, an algebraic relation between the roots a and b. Is the permutation R a symmetry? Well, if we check the definition above, R keeps a as it was and swaps b with c, so the condition $a + c^2 = 5$ must also hold. If it doesn't, R is definitely not a symmetry. If it does, you check any other valid algebraic relations among the roots, and if R passes *all* these tests, it is a symmetry.

Working out precisely which permutations are symmetries of a given equation is a difficult technical exercise. But there is one thing we can be sure of without doing any calculations at all. The collection of all symmetries of a given equation must be a subgroup of the group of all permutations of the roots.

Why? Suppose, for instance, that both P and R preserve all algebraic relations among the roots. If we take some relation and then apply R, we get a valid relation. If we then apply P, we again get a valid relation. But applying R and then P is the same as applying PR. So PR is a symmetry. In other words, the collection of symmetries has the group property.

This straightforward fact underlies the whole of Galois's work. It tells us that associated with any algebraic equation there is a group, its symmetry group—now called its *Galois group* to honor the inventor. And the Galois group of an equation is always a subgroup of the group of all permutations of the roots.

From this key insight a natural line of attack emerges. Understand which subgroups arise in which circumstances. In particular, if the equation can be solved by radicals, then the Galois group of the equation should reflect this fact in its internal structure. Then, given any equation, you just work out its Galois group and check whether it has the required structure, and you know whether it can be solved by radicals.

❋

Now Galois could recast the whole problem from a different viewpoint. Instead of building a tower with ladders and sacks, he grew a tree.

Not that he called it a tree, any more than Abel talked of Cardano Tower, but we can picture Galois's idea as a process that repeatedly branches from a central trunk. The trunk is the Galois group of the equation. The branches, twigs, and leaves are various subgroups.

Subgroups arise naturally as soon as we start thinking about how the symmetries of equations change when we start taking radicals. How does the group change? Galois showed that if we form a pth root, then the symmetry group must split into p distinct blocks, all the same size. (Here, as Abel noted, we can always assume p to be prime.) So, for example, a group of 15 permutations might split into five groups of 3, or three groups of 5. Crucially, the blocks have to satisfy some very precise conditions; in particular, one of them must form a subgroup in its own right of a special kind known as a "normal subgroup of index p." We can think of the trunk of the tree splitting into p smaller branches, one of which corresponds to the normal subgroup.

The normal subgroups of the group of all six permutations of three symbols are the entire group [I, U, V, P, Q, R], the subgroup [I, U, V], whose table we saw just now, and the subgroup with just one permutation, [I]. The other three subgroups, which contain two permutations, are not normal.

For instance, suppose we want to solve the general quintic. There are five roots, so the permutations involve five symbols. There are precisely 120 such permutations. The coefficients of the equation, being fully symmetric, have a group that contains all 120 of these. This group is the trunk of the tree. Each root, being totally asymmetric, has a group that contains just *one* permutation—the trivial one. So the tree has 120 leaves. Our aim is to join the trunk to the leaves by branches and twigs whose structure reflects the symmetry properties of the various quantities that arise if we start working out the bits and pieces of a formula for the roots, which we assume are expressed by radicals.

Suppose for the sake of argument that the first step in the formula is to adjoin a fifth root. Then the group of 120 permutations must split into five pieces, each containing 24 permutations. So the tree develops five branches. Technically, this branching must correspond to a normal subgroup of index 5.

However, Galois could prove, merely by calculating with permutations, that *there does not exist such a normal subgroup*.

Very well, perhaps the solution starts with, say, a seventh root. Then the 120 permutations must split into seven blocks of equal size—but they can't, because 120 is not divisible by 7. No seventh roots, then. In fact, no prime roots except 2, 3, and 5, because those are the prime factors of 120. And we've just ruled out 5.

A cube root to start with, then? Unfortunately not: the group of 120 permutations has no normal subgroup of index 3.

All that's left is a square root. Does the group of 120 permutations have a normal subgroup of index 2? Indeed it does, precisely one. It contains 60 permutations, and is called the *alternating group*. So by using Galois's theory of groups, we have established that any formula for solving a general quintic must start with a square root, leading to the alternating group. The first place where the trunk splits leads to just two branches. But there are 120 leaves, so the branches must split again. How do the branches split?

The prime divisors of 60 are also 2, 3, and 5. So each of our new branches must split into two, three, or five twigs. That is, we must either adjoin another square root, a cube root, or a fifth root. Moreover, this can be done if and only if the alternating group has a normal subgroup of index 2, 3, or 5.

But does it have such a normal subgroup? That is a question purely about permutations of five symbols. By analyzing such permutations, Galois was able to prove that the alternating group *has no normal subgroups at all* (except for the whole group and the trivial subgroup [*I*]). It is a "simple" group—one of the basic components out of which all groups can be constructed.

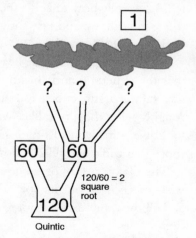

There are too few normal subgroups to connect the trunk to the leaves by means of splitting into a prime number of branches at each successive step. So the process of solving the quintic by radicals grinds to an abrupt halt after that first step of adjoining a square root. *There is nowhere else to go.* No tree

Galois's proof that the quintic is unsolvable.

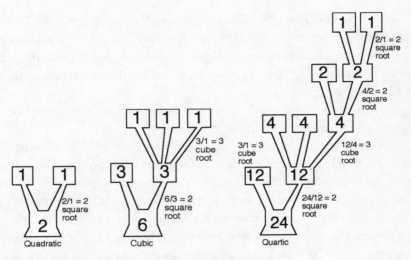

Using groups to solve the quadratic, cubic, and quartic.

can climb from the trunk all the way up to the leaves, and therefore there is no formula for the roots in terms of radicals.

The same idea works for equations of degree 6, 7, 8, 9—anything larger than 5. This leaves us wondering why the quadratic, cubic, and quartic *are* solvable. Why are degrees 2, 3, and 4 exceptional? In fact, group theory tells us exactly how to solve the quadratic, cubic, and quartic. I'll leave out the technicalities, and just show you the trees. They correspond precisely to the classical formulas.

Now we begin to see the beauty of Galois's idea. Not only does it prove that the general quintic has no radical solutions, it also explains why the general quadratic, cubic, and quartic *do* have radical solutions and tells us roughly what they look like. With extra work, it tells us *exactly* what they look like. Finally, it distinguishes those quintics that can be solved from those that can't, and tells us how to solve the ones that can.

The Galois group of an equation tells us everything we could possibly wish to know about its solutions. So why did Poisson, Cauchy, Lacroix, and all the other experts not leap with joy when they saw what Galois had done?

The Galois group has a terrible secret.

✳

The secret is this. The easiest way to work out the group of an equation is to use properties of its roots. But of course, the whole point is that we usually do not know what the roots are. Remember, we are trying to solve the equation, that is, to find its roots.

Suppose someone presents us with a specific quintic, say

$$x^5 - 6x + 3 = 0$$

or

$$x^5 + 15x + 12 = 0$$

and asks us to use Galois's methods to decide whether it can be solved by radicals. It seems a fair question.

The dreadful truth is that with the methods available to Galois, *there is no way to answer it.* We can assert that most likely, the associated group contains all 120 permutations—and if it does, then the equation can't be solved. But we don't know for sure that all 120 permutations actually occur. Maybe the five roots obey some special constraint. How can we tell?

Beautiful though it may be, Galois's theory has severe limitations. It works not with the coefficients but with the roots. In other words, it works with the unknowns, not with the knowns.

Today, you can go to a suitable mathematical website, input your equation, and it will calculate the Galois group. We now know that the first equation above is not solvable by radicals, but the second one *is* solvable. My point is not the computer, but that someone has discovered what steps should be taken to solve the problem. The great advance since Galois in this area was working out how to compute the Galois group of any given equation.

Galois possessed no such technique. It would take another century before routine calculation of the Galois group became feasible. But the absence of this technique let Cauchy and Poisson off the hook. They could complain, with complete justification, that Galois's ideas did not solve the problem of deciding when a *given* equation could be solved by radicals.

What they failed to appreciate was that his method did solve a slightly different problem: to work out which properties of the *roots* made an equation solvable. That problem had an elegant—and deep—answer. The problem they wanted him to solve . . . well, there is no reason to expect a neat answer. There just isn't a tidy way to classify the solvable equations in terms of easily computed properties of their coefficients.

So far, the interpretation of groups as symmetries has been somewhat metaphorical. Now we need to make it more literal, and that step requires a more geometric point of view. Galois's successors quickly realized that the relation between groups and symmetry is much easier to understand in the context of geometry. In fact, this is how the subject is usually introduced to students.

To get a feeling for this relationship, we'll take a quick look at my favorite group: the symmetry group of an equilateral triangle. And we'll finally address a very basic question: What, exactly, *is* symmetry?

Before Galois, all answers to this question were rather vague, handwavy things, with appeals to features like elegance of proportion. This is not a concept you can do sensible mathematics with. After Galois—and after a short period during which the world of mathematics sorted out the general ideas behind his very specific application—there was a simple and unequivocal answer. First, the word "symmetry" has to be reinterpreted as "*a* symmetry." Objects do not possess symmetry alone; they often possess many different symmetries.

What, then, is a symmetry? A symmetry of some mathematical object is a transformation that preserves the object's structure. I'll unpack this definition in a moment, but the first point to observe is that a symmetry is a process rather than a thing. Galois's symmetries are permutations (of the roots of an equation), and a permutation is *a way to rearrange things*. It is not, strictly speaking, the rearrangement itself; it is the rule you apply to get the rearrangement. Not the dish but the recipe.

This distinction may sound like splitting hairs, but it is fundamental to the whole enterprise.

There are three key words in the definition of a symmetry: "transformation," "structure," and "preserve." Let me explain them using the example of an equilateral triangle. Such a triangle is defined as having all three sides the same length and all three angles the same size, namely 60°. These features make it difficult to distinguish one side from another; phrases like "the longest side" don't tell us anything. The angles are also indistinguishable. As we now see, the inability to distinguish one side from another or one angle from another is a consequence of the symmetries of the equilateral triangle. In fact, it is what defines those symmetries.

Let's consider those three words in turn.

Transformation: We are allowed to do things to our triangle. In principle there are lots of things we might do: bend it, turn it through some angle, crumple it up, stretch it like elastic, paint it pink. Our choice here is more limited, however, by the second word.

Structure: The structure of our triangle consists of the mathematical features that are considered significant. The structure of a triangle includes such things as "it has three sides," "the sides are straight," "one side has length 7.32 inches," "it sits in the plane at this location," and so on. (In other branches of mathematics, the significant features may be different. In topology, for instance, what matters is that the triangle forms a single closed path, but its three corners and the straightness of its edges are no longer important.)

Preserve: The structure of the transformed object must match that of the original. The transformed triangle must also have three sides, so crumpling it is ruled out. The sides must remain straight, so bending it is not permitted. One side must still have length 7.32 inches, so stretching the triangle is forbidden. The location must be the same, so sliding it ten feet sideways is disallowed.

The color is not explicitly mentioned as structure, so painting the triangle pink is irrelevant. It's not exactly ruled out; it just makes no difference for geometric purposes.

Turning the triangle through some angle, however, does preserve at least some of the structure. If you make an equilateral triangle out of cardboard, set it on the table, and then rotate it, it still looks like a triangle. It has three sides, they are still straight, their lengths haven't changed. But the location of the triangle in the plane may still look different, depending on the angle through which you rotate it.

If I turn the triangle through a right angle, for instance, the result looks different. The sides point in different directions. If you covered your eyes while I turned the triangle, you would know when you opened them again that I had moved it.

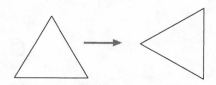

Rotation through a right angle is not a symmetry of the equilateral triangle.

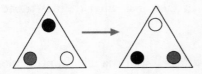

Rotation through 120° is a symmetry of the equilateral triangle.

But if I turned the triangle through 120°, you wouldn't be able to see any difference between "before" and "after." To show you what I mean, I will secretly mark the corners with different types of dots, so we can see where it moves to. These dots are for reference only and are not part of the structure that is preserved. If you can't see the dots, if the triangle is as featureless as any well-behaved Euclidean object, then the turned triangle looks the same as the original.

In other words, "rotate by 120°" is a symmetry of the equilateral triangle. It is a transformation ("rotate") that preserves the structure (shape and location).

It turns out that an equilateral triangle has precisely six different symmetries. Another is "rotate by 240°." Three more are reflections, which turn the triangle over so that one corner remains fixed and the other two exchange positions. What is the sixth symmetry? *Do nothing.* Leave the triangle alone. This is trivial, but it fits the definition of symmetry. In fact, this transformation fits the definition of symmetry no matter what object we consider or what structure we want preserved. If you do nothing, then nothing changes.

This trivial symmetry is called the *identity.* It may seem insignificant, but if we leave it out, the math gets very messy. It's like trying to do addition without the number zero or multiplication without the number one. If we keep the identity in, everything stays neat and tidy.

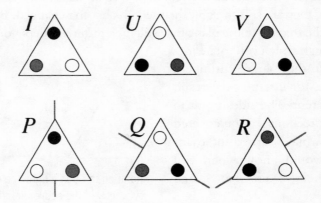

The six symmetries of the equilateral triangle.

For the equilateral triangle, you can think of the identity as rotation through 0°. On the previous page are the results of applying the six symmetries to our equilateral triangle. They are precisely the six different ways that you can pick up a triangle made of cardboard and lay it down within its original outlines. The dotted lines show where to put the mirror to obtain the required reflection.

Now I want to convince you that symmetries are a part of algebra. So I will do what any algebraist would do: express everything in terms of symbols. We will name the six symmetries I, U, V, P, Q, R according to the picture above. The identity is I; the other two rotations are U and V; the three reflections are P, Q, and R. These are the same symbols that I used earlier for permutations of the roots of a cubic. There is a reason for this duplication, which will shortly emerge.

Galois made great play of the "group property" of his permutations. If you perform any two in turn, you get another one. This provides a big hint about what we should do with our six symmetries. We should "multiply" them in pairs and see what happens. Recall the convention: if X and Y are two symmetry transformations, then the product XY is what happens when we first do Y, then X.

Suppose, for instance, that we want to work out VU. This means that first we apply U to the triangle, then V. Well, U rotates it through 120°, and V then rotates the resulting triangle through 240°. So VU rotates it through 120° + 240° = 360°.

Oops, we forgot to include that.

No we didn't. If you rotate a triangle through 360°, everything ends up exactly where it started. And in group theory it is the end result that matters, not the route taken to get there. In the language of symmetries, two symmetries are considered to be the same if they have the same final effect on the object. Since VU has the same effect as the identity, we conclude that $VU = I$.

For a second example, what does UQ do? The transformations go like this:

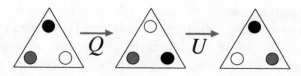

How to multiply symmetries.

We recognize the end result: it is P. So $UQ = P$.

With our six symmetries we can form 36 products, and the calculations can be captured in a multiplication table. It is exactly the same table that we obtained for the six permutations of the roots of a cubic.

This apparent coincidence turns out to be an example of one of the most powerful techniques in the whole of group theory. It originated in the work of the French mathematician Camille Jordan, who arguably turned group theory into a subject in its own right, rather than just a method for analyzing the solution of equations by radicals.

Around 1870, Jordan drew attention to what is now called "representation theory." To Galois, groups were composed of permutations—ways to shuffle symbols. Jordan started thinking about ways to shuffle more complicated spaces. Among the most basic spaces in mathematics are multidimensional spaces, and their most important feature is the existence of straight lines. The natural way to transform such spaces is *to keep straight lines straight.* No bending, no twisting. There are many transformations of this kind—rotations, reflections, changes of scale. They are called "linear" transformations.

The English lawyer-mathematician Arthur Cayley discovered that any linear transformation can be associated with a *matrix*—a square table of numbers. Any linear transformation of three-dimensional space, for example, can be specified by writing down a 3-by-3 table of real numbers. So transformations can be reduced to algebraic computations.

Representation theory lets you start with a group that does not consist of linear transformations and replace it with one that does. The advantage of converting the group to a group of matrices is that matrix algebra is very deep and powerful, and Jordan was the first to see this.

Let's look at the symmetries of the triangle from Jordan's point of view. Instead of placing shaded dots in the corners of the triangle, I will place the symbols *a, b, c*, corresponding to the roots of the general cubic. It then becomes obvious that each symmetry of the triangle also permutes these symbols. For example, the rotation *U* sends *abc* to *cab*.

The six symmetries of the triangle correspond naturally to the six permutations of the roots *a, b, c*. Moreover, the product of two symmetries corresponds to the product of the corresponding permutations. But rota-

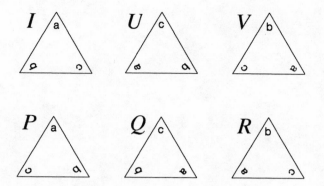

How symmetries of the equilateral triangle correspond to permutations.

tions and reflections in the plane are linear transformations—they preserve straight lines. So we have reinterpreted the permutation group—*represented* it—as a group of linear transformations, or equivalently as a group of matrices. This idea was to have profound consequences for both mathematics and physics.

8

THE MEDIOCRE ENGINEER
AND THE TRANSCENDENT PROFESSOR

No longer was symmetry some vague impression of regularity or an artistic feeling of elegance and beauty. It was a clear mathematical concept with a rigorous logical definition. You could calculate with symmetries and prove theorems about them. A new subject was born: *group theory*. Humanity's quest for symmetry had reached a turning point. The admission fee for this advance was a willingness to think more conceptually. The concept of a group was an abstract one, several stages removed from the traditional raw materials of numbers and geometrical shapes.

Groups had already proved their worth by resolving an age-old conundrum, the solvability of the quintic. It soon became clear that the same circle of ideas disposed of several other age-old problems. You didn't always need group theory as such, but you needed to think like Abel, Galois, and their successors. And even when you thought you weren't using groups, they often lurked in the background.

Among the unsolved problems the Greek geometers bequeathed to posterity, three had become notorious: the problems of trisecting the angle, duplicating the cube, and squaring the circle. Even today, trisection and circle-squaring attract the attention of numerous amateurs, who seem not to have grasped that when mathematicians use the word "impossible," they mean it. Duplicating the cube seems not to have the same allure.

These three are often referred to as the "three problems of antiquity," but this phrase exaggerates their importance. It makes them appear to be on a par with major historical puzzles such as Fermat's Last Theorem, which went unanswered for more than 350 years. But this puzzle was explicitly recognized as an unsolved problem, and it is possible to identify the precise point in the mathematical literature where it was first posed. All mathematicians were aware not just of the problem but the presumed answer—and who had first asked the question.

The Greek problems are not like that. You won't find them listed in Euclid as unsolved problems that need attention. They exist mainly by default: they are obvious extensions of positive results, but for some reason Euclid avoided them. Why? Because no one knew how to solve them. Did it occur to the Greeks that they might not *have* solutions? If so, no one made much fuss. It undoubtedly occurred to people like Archimedes that no straightedge-and-compass solutions existed, because he developed alternative techniques, but there is no evidence that Archimedes considered the issue of constructibility important in its own right.

Later it became important. The lack of solutions to these problems pointed to major gaps in humanity's understanding of geometry and algebra; they gained currency as "folklore" problems, known to the professionals through a kind of cultural osmosis. By the time they were solved, they had taken on an aura of historical and mathematical significance. Their solutions were seen as major breakthroughs—especially squaring the circle. And in all three cases, the answer was the same: "It can't be done." Not with the traditional tools of straightedge and compass.

This may seem rather negative. In most walks of life, people seek to answer questions or overcome difficulties by whatever means comes to hand. If a tall building cannot be constructed from bricks and mortar, engineers use steel frames and reinforced concrete. No one gains fame by proving that bricks are not up to the job.

Mathematics is not quite like that. The limitations of the tools are often just as important as what they can accomplish. The importance of a mathematical question often depends not on the answer as such, but on why the answer is correct. So it was for the three problems of antiquity.

⁎

The scourge of trisectors everywhere was born in Paris in 1814, and his name was Pierre Laurent Wantzel. His father was first an army officer and later professor of applied mathematics at the École Speciale du Commerce. Pierre was precocious; Adhémard Jean Claude Barré de Saint-Venant, who knew Wantzel, wrote that the boy showed "a marvelous aptitude for mathematics, a subject about which he read with extreme interest. He soon surpassed even his master, who sent for the young Wantzel, at age nine, when he encountered a difficult surveying problem."

In 1828, Pierre successfully applied to enter the Collège Charlemagne. He won first prize in both French and Latin in 1831, and he came in first in both the entrance examination for the École Polytechnique and that for the science section of what is now the École Normale, which no one had ever done before. He was interested in just about everything—mathematics, music, philosophy, history—and he liked nothing better than a good, hard-fought debate.

In 1834, he turned his mind to engineering, attending the École des Ponts et Chaussées. But soon he was confessing to his friends that he would be "but a mediocre engineer," decided that he really wanted to teach mathematics, and took a leave of absence. The switch worked: he became a lecturer in analysis at the École Polytechnique in 1838, and by 1841 he was also a professor of applied mechanics at his old engineering school. Saint-Venant tells us that Pierre "usually worked during the evening, not going to bed until late in the night, then reading, and got but a few hours of agitated sleep, alternatively abusing coffee and opium, taking his meals, until his marriage, at odd and irregular hours." The marriage was to his former Latin coach's daughter.

Wantzel studied the works of Ruffini, Abel, Galois, and Gauss, developing a strong interest in the theory of equations. In 1837 his paper "On the means of ascertaining whether a geometric problem can be solved with straightedge and compass" appeared in Liouville's *Journal de Mathématiques Pures et Appliquées*. It took up the story of constructibility where Gauss had left off. He died in 1848 at the age of 33—probably as a result of overwork from an excess of teaching and administrative duties.

On the questions of trisection and duplicating the cube, Wantzel's impossibility proofs resemble Gauss's epic work on regular polygons, but are much easier. I'll start with the duplication of the cube, where the issues are very transparent. Does there exist a straightedge-and-compass construction for a line of length $\sqrt[3]{2}$?

Gauss's analysis of regular polygons is based on the idea that any geometric construction boils down to solving a series of quadratic equations. He pretty much takes this for granted, because it follows algebraically from properties of lines and circles. Some reasonably easy algebra implies that the "minimum polynomial" of any constructible quantity—the simplest equation that it satisfies—has degree equal to a power of two. That equation may be linear, quadratic, quartic, octic (degree 8), of degree 16, 32, 64, . . . but whatever the degree is, it is a power of two.

On the other hand, $\sqrt[3]{2}$ satisfies the cubic equation $x^3 - 2 = 0$, and this is its minimum polynomial. The degree is 3, which is not a power of 2. Therefore the assumption that the cube can be duplicated using straightedge and compass leads, by impeccable logic, to the conclusion that 3 is a power of 2. This is obviously not true. By *reductio ad absurdum,* therefore, no such construction can exist.

Trisection of the angle is impossible for a similar reason, but the proof is slightly more involved.

First, *some* angles can be trisected exactly. A good example is 180°, which trisects to 60°, an angle that we can construct by making a regular hexagon. So the impossibility proof begins by picking some other angle and proving that this choice cannot be trisected. The simplest angle to pick is 60° itself. One-third of that is 20°, and we will show that 20° cannot be constructed using straightedge and compass.

This is a sobering thought. Look at a protractor, an instrument for measuring angles. Confidently marked on it are angles of 10°, 20°, and so on. But those angles are not exact—for a start, the inked lines have thickness. We can make an angle of 20° that's good enough for an architectural or engineering drawing. But we can't construct a perfect 20° angle using Euclidean methods—that's what we plan to prove.

The key to this puzzle is trigonometry, the quantitative study of angles. Suppose that we start with a hexagon inscribed in a circle of radius 1. Then we can find a 60° angle, and if we could trisect it, we could construct the bold line in the figure (next page).

Suppose this line has length x. Trigonometry informs us that x satisfies the equation $8x^3 - 6x - 1 = 0$. As in the problem of duplicating the cube, this is cubic, and again it is the minimum polynomial of x. But if x is constructible then the degree of its minimum polynomial must be a power of

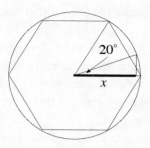

Trisecting an angle of 60° is
equivalent to constructing the
length marked x.

2. Same contradiction, same conclusion: the proposed construction is impossible.

The way I have presented these proofs conceals a deeper structure, and from a more abstract perspective Wantzel's solutions of these two problems of antiquity both boil down to symmetry arguments: the Galois groups of the equations that correspond to the geometry have the wrong structure for straightedge-and-compass constructions. Wantzel was well aware of Galois groups, and in 1845 he developed a new proof that some algebraic equations cannot be solved by radicals. The proof followed Ruffini and Abel closely, but simplified and clarified the ideas. In the introduction Wantzel states,

> Although [Abel's] proof is finally correct, it is presented in a form too complicated and so vague that it is not generally accepted. Many years previous, Ruffini . . . had treated the same question in a manner much vaguer still . . . In meditating on the researches of these two mathematicians . . . we have arrived at a form of proof which appears so strict as to remove all doubt on this important part of the theory of equations.

The sole remaining problem of antiquity was squaring the circle, a task that amounts to constructing a line of length *exactly* equal to π. Proving this construction impossible turned out to be much more difficult. Why? Because instead of π having a minimum polynomial of the wrong degree, it turned out to have no minimum polynomial *at all*. There is no polynomial equation with rational coefficients having a root equal to π. You can come as close as you like, but you can never get exactly π.

The mathematicians of the nineteenth century realized that the distinction between rational and irrational numbers could profitably be refined. There were different kinds of irrational. Relatively "tame" irrationals like $\sqrt{2}$ could not be represented as exact fractions, that is, as rational numbers, but they could be represented *in terms of* rational numbers. They satisfied equations whose coefficients were rational numbers—in this case, $x^2 - 2 = 0$. Such numbers were said to be "algebraic."

But mathematicians realized that in principle there might exist irrational numbers that were *not* algebraic and whose link to the rationals was far more indirect than that for the algebraic numbers. They *transcended* the rational realm altogether.

The first question was, do such "transcendental" numbers actually exist? The Greeks supposed that all numbers might be rational until Hippasus disillusioned them, and Pythagoras allegedly was so incensed that he drowned the messenger. (More likely, Hippasus was just expelled from the Pythagorean cult.) The mathematicians of the nineteenth century were aware that any belief that all numbers are algebraic was equally likely to lead to tragedy, but for many years they lacked a Hippasus. All they had to do was to prove that some specific real number—π was a plausible candidate—is not algebraic. But it's difficult enough to prove that some number, π, for example, is irrational, and for that all you have to show is the nonexistence of any pair of integers such that one divided by the other gives you π. To prove that a number is not algebraic, you have to replace these hypothetical integers by all possible equations, of any degree whatsoever, and then derive a contradiction. It gets messy.

The first significant progress was made by the German mathematician and astronomer Johann Lambert in 1768. In a paper on transcendental numbers, he proved that π is irrational, and his method paved the way to everything that followed. It made essential use of ideas from calculus, notably the concept of an "integral." (The integral of any given function is a function whose rate of change yields the original function.) Starting from the assumption that π is equal to some exact fraction, Lambert proposed to calculate a fairly complicated integral that he had invented for just this purpose, which involved not just polynomials but trigonometric functions. There are two distinct ways to calculate this integral. One of them gives the answer zero. The other proves that the answer is *not* zero.

If π were not a fraction, then neither method would apply, so no problems would arise. But if π were a fraction, zero would have to be different from itself. No way.

The details of Lambert's proof are technical, but how it works is very informative. To get started, he had to relate π to something simpler, and trigonometry came to his rescue. The next problem was to fix things up so that something special would happen if π were rational. This was where the polynomial bit came in, along with the clever idea of forming an integral. After that, the proof was a matter of comparing two distinct methods for computing the integral, and showing that they gave different answers. That bit was messy and technical, but routine for experts.

Lambert's proof was a major step forward, but plenty of irrational numbers can be constructed, the most obvious being $\sqrt{2}$, the diagonal of a unit square. So proving π irrational did not prove that it was unconstructible. It meant there was no longer any point in trying to find an exact fraction for π, but that was a different issue altogether.

At this point, mathematicians faced an unusual dilemma. They had made a distinction between algebraic numbers and transcendental ones, and they believed it would be important. But they still did not know whether any transcendental numbers existed. In practical terms, the supposed distinction might be meaningless.

It took until 1844 to prove the existence of transcendentals. The breakthrough was made by Liouville, who had previously salvaged Galois's work from the academic rubbish heap. Now Liouville managed to invent a transcendental number. It looked like this:

$$0.110001000000000000000001000...,$$

where longer and longer sequences of 0's are separated by isolated 1's. The important point is that the lengths of the blocks of zeros have to increase very rapidly.

Numbers of this kind are "almost" rational. There exist unusually good rational approximations—basically thanks to those blocks of zeros. The long block above, for instance, with 17 consecutive zeros, implies that what comes before it—0.110001—is a much better approximation to Liouville's number than you might expect of a random decimal fraction. And 0.110001, like any finite decimal, is rational: it is equal to $\frac{110001}{1000000}$. Instead of being accurate to six decimal places, it is accurate to 23 decimal places. The next nonzero digit is a 1 in the 24th place.

Liouville had realized that algebraic numbers, other than rational ones, are always rather *badly* approximated by rationals. Not only are such numbers irrational; to get a good rational approximation you have to use very big numbers in any fraction that gets close. So Liouville deliberately defined his number to have extraordinarily good rational approximations, much too good for it to be algebraic. Therefore it had to be transcendental.

The only criticism we can direct against this clever idea is that Liouville's number is very artificial. It has no evident connection with anything else in mathematics. It is plucked from thin air for the sole reason that it can be very well approximated by rationals. No one would care about it at all save for that one remarkable feature: it is provably transcendental. So now mathematicians knew that transcendentals did exist.

Whether *interesting* transcendentals existed was another matter, but at least the theory of transcendental numbers had some content. Now the task was to provide interesting content. Above all, is π transcendental? If it were, that would knock the old squaring-the-circle problem on the head. All constructible numbers are algebraic, so no transcendental is constructible. If π is transcendental, it is impossible to square the circle.

The number π is justly famous because of its connections with circles and spheres. Still, mathematics contains other remarkable numbers, and the most important—probably even more important than π—is known as e. Its numerical value is approximately 2.71828, and like π it is irrational. This number arose in 1618, in the early days of logarithms; it determines the correct interest rate if compound interest is applied over ever-shorter intervals. It was called b in a letter Leibniz wrote to Huygens in 1690. The symbol e was introduced by Euler in 1727, and it appeared in print in his *Mechanics* of 1736.

By using complex numbers, Euler discovered a remarkable relation between e and π, often considered the most beautiful formula in mathematics. He proved that $e^{i\pi} = -1$. (This formula does have an intuitive explanation, but it involves differential equations.) After Liouville's discovery, the next step to the proof that π is transcendental took a further 29 years, and it applied to the number *e*. In 1873 the French mathematician Charles Hermite proved that *e* is transcendental. Hermite's career has remarkable parallels with that of Galois—he went to Louis-le-Grand, was taught by Richard, tried to prove that the quintic is unsolvable, and wanted to study at the École Polytechnique. But unlike Galois he got in—by the skin of his teeth.

One of Hermite's students, the famous mathematician Henri Poincaré, observed that Hermite's mind worked in strange ways: "To call Hermite a logician! Nothing can appear to me more contrary to the truth. Methods always seemed to be born in his mind in some mysterious way." This originality served Hermite well in his proof that *e* is transcendental. The proof was an elaborate generalization of Lambert's proof that π is irrational. It also employed calculus; it evaluated an integral in two ways; and if *e* were algebraic, those two answers would be different: one equal to zero, one nonzero. The difficult step was to find the right integral to compute.

The actual proof occupies about two printed pages. But what a wonderful two pages! You could search for a lifetime and not discover the right choice of integral.

The number *e* is at least a "natural" object of mathematical study. It crops up all over mathematics and is absolutely vital to complex analysis and the theory of differential equations. Although Hermite had not cracked the problem of π, he had at least improved on Liouville's rather artificial example. Now mathematicians knew that the everyday operations of mathematics could throw up entirely reasonable numbers that turned out to be transcendental. Soon a successor would use Hermite's ideas to prove that one of those numbers was π.

Carl Louis Ferdinand von Lindemann was born in 1852, the son of a language teacher, Ferdinand Lindemann, and the headmaster's daughter, Emilie Crusius. Ferdinand changed jobs, becoming the director of a gasworks.

Like many students in late-nineteenth-century Germany, Lindemann Jr. moved from one university to another—Göttingen, Erlangen, Munich. At Erlangen he took a PhD on non-Euclidean geometry under the supervision of Felix Klein. He traveled abroad, to Oxford and Cambridge, and then to Paris, where he met Hermite. On obtaining his habilitation in

1879, he obtained a professorship at the University of Freiburg. Four years later he moved to the University of Königsberg, where he met and married Elizabeth Küssner, a teacher's daughter who worked as an actress. Ten years after that, he became a full professor at the University of Munich.

In 1882, halfway between his trip to Paris and his appointment to Königsberg, Lindemann figured out how to extend Hermite's method to prove the transcendence of π, and became famous. Some historians believe that Lindemann just got lucky—that he was a bit of a hack who blundered across the right extension of Hermite's magnificent idea. But as the golfer Gary Player once remarked, "The better I play, the luckier I get." So, most likely, was it with Lindemann. If *anyone* could get lucky, why didn't Hermite?

Later, Lindemann turned to mathematical physics, investigating the electron. His most famous research student was David Hilbert.

Lindemann's proof of the transcendence of π used the method pioneered by Lambert and developed by Hermite: write down a suitable integral, calculate it two ways, and show that if π is algebraic the answers disagree. The integral was very closely related to the one used by Hermite, but even more complicated. The connection between e and π, in fact, was the beautiful relationship discovered by Euler. If π were algebraic, then e would have to have some new and surprising properties—analogous to, but differing from, being algebraic. The core of Lindemann's proof is about e, not about π.

With Lindemann's proof, this chapter of mathematics reached its first truly significant conclusion. That it was impossible to square the circle was barely a sideshow. Much more important was that mathematicians knew why. Now they could go on to develop the theory of transcendental numbers, which today is an active—and fiendishly difficult—area of research. Even the most obvious and plausible conjectures about transcendental numbers remain mostly unanswered.

※

Armed with the insights of Abel and Galois, we can revisit the problem of constructing regular polygons. For which numbers n is the regular n-gon constructible with straightedge and compass? The answer is extraordinary.

In the *Disquisitiones Arithmeticae,* Gauss stated necessary and sufficient conditions on the integer $n,$ but he proved only their sufficiency. He claimed to possess a proof that these same conditions are also necessary, but—like much of his work—he never published it. Gauss had actually done the hard part, and it was Wantzel who filled in the missing details in his 1837 paper.

To motivate Gauss's answer, we briefly review the regular 17-gon. What is it about the number 17 that makes the regular 17-sided polygon constructible? Why is this not the case for numbers like 11 or 13?

Notice that all three numbers here are primes. It is easy to show that if a regular n-gon is constructible, then so is the regular p-gon for every prime p dividing n. Just take every n/pth corner. For example, taking every third vertex of a regular 15-gon yields a regular 5-gon. So it makes sense to think about prime numbers of sides, and to use the results for the primes to work your way toward a complete solution.

The number 17 is prime, so that's a good start. Gauss's analysis, reformulated in more modern terms, is based on the fact that the solutions of the equation $x^{17} - 1 = 0$ form the vertices of a regular 17-gon in the complex plane. There is one obvious root, $x = 1$. The other 16 are the roots of a polynomial of degree 16, which can be shown to be $x^{16} + x^{15} + x^{14} + \ldots x^2 + x + 1 = 0$. The 17-gon is constructed by solving a series of quadratic equations, and it turns out that this is possible because 16 is a power of 2. It equals 2^4.

More generally, the same line of argument proves that when p is an odd prime, the regular p-gon is constructible *if and only if $p - 1$* is a power of 2. Such odd primes are called Fermat primes because Fermat was the first to investigate them. The Greeks knew of constructions for the regular 3-gon and the regular 5-gon. Observe that $3 - 1 = 2$, and $5 - 1 = 4$, both powers of 2. So the Greek results are consistent with Gauss's criterion, and 3 and 5 are the first two Fermat primes. On the other hand, $7 - 1 = 6$, not a power of two, so the regular 7-gon is not constructible.

A bit of extra work leads to Gauss's characterization: the regular n-gon is constructible if and only if n is a power of two, or a power of two multiplied by *distinct* Fermat primes.

This leaves the question, what are the Fermat primes? The next one after 3 and 5 is Gauss's discovery, 17. The next is 257, followed by the rather large number 65,537. These are the only known Fermat primes. It has never been proved that no further Fermat primes exist—but it has also never been proved that they don't. For all we know, there might be some absolutely gigantic Fermat prime not yet known to humanity. At the current state of knowledge this number is at least $2^{33554432} + 1$, and indeed that might be the next Fermat prime. (The exponent 33554432 is itself a power of 2, namely 2^{25}. All Fermat primes are one greater than two raised to the power of *a power of* two.) This number has more than ten million digits. Even after Gauss's great discoveries, we still do not know for sure exactly which regular polygons are constructible, but the only gap in our knowledge is the possible existence of very large Fermat primes.

Although Gauss proved that the 17-gon is constructible, he did not actually describe the construction as such, though he did remark that the main point is to construct a line of length

$$\frac{1}{16}\left[-1 + \sqrt{17} + \sqrt{34 - 2\sqrt{17}} + \sqrt{68 + 12\sqrt{17} - 16\sqrt{34 + 2\sqrt{17}} - 2(1 - \sqrt{17})(\sqrt{34 - 2\sqrt{17}})}\right]$$

The formula that determines Gauss's construction of the regular 17-gon.

Since square roots are always constructible, the required construction is implicit in this remarkable number. The first explicit construction was devised by Ulrich von Huguenin in 1803. H. W. Richmond found a simpler version in 1893.

In 1832, F. J. Richelot published a series of papers constructing the regular 257-gon, under the title "De resolutione algebraica aequationis $x^{257} = 1$, sive de divisione circuli per bisectionem anguli septies repetitam in partes 257 inter se aequales commentatio coronata," which is even more impressive than the number of sides of his polygon.

There is an apocryphal tale that an overzealous PhD student was assigned the construction of the 65537-gon as a thesis project, and reappeared with it twenty years later. The truth is almost as bizarre: J. Hermes, of the University of Lingen, devoted ten years to the task, finishing in 1894, and his unpublished work is preserved at the University of Göttingen. Unfortunately, John Horton Conway, perhaps the only mathematician to have looked at these documents in modern times, doubts that the work is correct.

9

THE DRUNKEN VANDAL

William Rowan Hamilton was the greatest mathematician Ireland has ever produced. He was born at the stroke of midnight between 3 and 4 August 1805, and never quite made up his mind which of those dates was his birthday. Mostly he settled for the 3rd, but his tombstone bears the date 4 August, because he switched to that date in later life for sentimental reasons. He was a brilliant linguist, a mathematical genius, and an alcoholic. He set out to invent an algebra of three dimensions but realized, in a flash of intuition that caused him to vandalize a bridge, that he would have to settle for four dimensions instead. He forever changed the human view of algebra, space, and time.

William was born into a wealthy family, the third son of Archibald Hamilton, a lawyer with a sound head for business. William also had a sister, Eliza. His father was partial to the odd glass or three, which made him good company for a while but a growing embarrassment as the evening wore on. Archibald was articulate, intelligent, and religious, and he passed on all of his significant traits, alcohol and all, to his youngest son. William's mother, Sarah Hutton, was even more intelligent than her husband, and came from a family of intellectual distinction, but her influence on young William, other than through her genes, was cut short when the father packed the boy off to be tutored by his uncle James at the age of three. James was a curate and an accomplished linguist, and his interests determined the main direction of William's education.

The results were impressive but obsessively narrow. By the age of five, William was fluent in Greek, Latin, and Hebrew. By eight he could speak French and Italian. Two years later he had added Arabic and Sanskrit; then Persian, Syrian, Hindu, Malay, Mahratti, and Bengali. Attempts to teach the lad Chinese were stymied by a lack of suitable texts. James complained that "it cost me a large sum to supply him from London, but I hope the money was well expended."

The mathematician and quasi-historian Eric Temple Bell ("quasi" because he never let an awkward fact get in the way of a good story) asked, "What was it all *for*?"

Fortunately for science and mathematics, William was saved from a life of mastering ever more of the world's thousands of languages when he came into contact with the American calculating prodigy Zerah Colburn. Colburn was one of those strange people who resemble a human pocket calculator; he had a talent for rapid, accurate computation. If you asked Colburn for the cube root of 1,860,867, he would reply "123" without pausing for breath.

This talent is distinct from mathematical ability, just as a facility for spelling does not make a good novelist. Except for Gauss, who left numerous big calculations in his notebooks and manuscripts, very few of the great mathematicians were lightning calculators. The rest were competent calculators—in those days you had to be—but no better than a qualified accountant. Even today, computers have not completely rendered pencil-and-paper calculations, or mental ones, obsolete; you can often gain insight into a mathematical problem by doing the calculations by hand and watching the symbols shuffle themselves around. But given the right software, much of it written by mathematicians, anyone with an hour's training can knock the socks off the likes of Colburn.

None of this will make you remotely resemble Gauss.

Colburn did not fully understand the tricks and short cuts he employed, though he was aware that memory played a big part. He was introduced to Hamilton in the expectation that the youthful genius would be able to shed light on these mysterious techniques. William did that and even came up with improvements. By the time Colburn departed, Hamilton had finally found a topic worthy of his astonishing brainpower.

By the age of seventeen, Hamilton had read many of the works of the mathematical masters, and knew enough mathematical astronomy to calculate eclipses. He still spent more time on the classics than on mathematics, but the latter had become his true passion. Soon he was making new discoveries. Just as the 19-year-old Gauss discovered the construction of a regular 17-gon, so the young Hamilton made an equally unprecedented breakthrough, an analogy—mathematically, an identity—between mechanics and optics, the science of light. He first alluded to these ideas in a cryptic letter to his sister Eliza, but we can be fairly certain of their nature from a subsequent letter to his cousin Arthur.

The discovery was amazing. Mechanics is the study of moving bodies—cannonballs traveling in a parabolic arc, pendulums swinging regularly from side to side, and planets moving in ellipses round the Sun. Optics is about the geometry of light rays, reflection and refraction, rainbows and prisms and telescope lenses. That they were connected was a surprise; that they were *the same* was unbelievable.

It was also true. And it led directly to the formal setting used today by mathematicians and mathematical physicists, not just in mechanics and optics but in quantum theory too: Hamiltonian systems. Their main feature is that they derive the equations of motion of a mechanical system from a single quantity, the total energy, now called the *Hamiltonian* of the system. The resulting equations involve not just the positions of the parts of the system but how fast they are moving—the *momentum* of the system. Finally, the equations have the beautiful feature that they do not depend on the choice of coordinates. Beauty is truth, at least in mathematics. And here the physics is both beautiful and true.

Hamilton was luckier than either Abel or Galois in that his unusual talents were widely recognized from early childhood. So it was no surprise when, in 1823, he gained admission to Ireland's leading university, Trinity College Dublin. Nor was it a surprise to find him at the head of a field of one hundred candidates. At Trinity, he took all the prizes. More importantly, he finished the first volume of his masterpiece on optics.

In the spring of 1825, Hamilton discovered the attractions of the fairer sex, in the form of one Catherine Disney. Unwisely, he confined his attentions to writing poems, and his would-be love promptly married a wealthy clergyman, fifteen years her senior, who had a less literary approach to fair damsels. Hamilton was devastated; despite being staunchly religious, he thought about drowning himself, a mortal sin. Second thoughts prevailed, and he consoled himself by pouring out his frustrations in yet another poem.

Hamilton loved poetry, and his circle of friends included leading members of the literati. William Wordsworth became a close friend; he also spent time with Samuel Taylor Coleridge and various other writers and poets. Wordsworth performed the valuable service of gently pointing out to Hamilton that his talents did not lie in poetry: "You send me showers of verses which I receive with much pleasure . . . yet have we fears that this employment may seduce you from the path of science . . . I do venture to submit to your consideration, whether the poetical parts of your nature would not find a field more favorable to their nature in the regions of prose . . ."

Hamilton responded that his true poetry was his mathematics, and wisely turned to science. In 1827, while still an undergraduate, he was unanimously elected Professor of Astronomy at Trinity after the incumbent, John Brinkley, resigned to become Bishop of Cloyne. Hamilton started with a bang by publishing his book on optics—an entirely valid topic for an astronomer since it underlay the design of most astronomical instruments.

The link to mechanics was present only in embryonic form. The book's main focus, so to speak, was on the geometry of light rays—how they change direction when reflected in a mirror or refracted in a lens. "Ray optics" later gave way to "wave optics," which recognized that light is a wave. Waves possess all sorts of extra properties, notably diffraction. Interference among waves can soften the edges of a projected image and even make light seem to bend around corners, a trick forbidden to rays.

The geometry of light rays was not a new topic; it had been studied extensively by earlier mathematicians, right back to Fermat, indeed, back to the Greek philosopher Aristotle. Now Hamilton did for optics what Legendre had famously done for mechanics: he got rid of the geometry and replaced it by algebra and analysis. Specifically, he replaced overt geometrical reasoning, based on diagrams, with symbolic calculations.

This was a major advance, because it replaced imprecise pictures with rigorous analysis. Later mathematicians made strenuous efforts to reverse Hamilton's path and reintroduce visual thinking. But by then, the formal algebraic stance had become part and parcel of mathematical thought, a natural companion to more overtly visual arguments. The wheel of fashion had come full circle but on a higher level, like a spiral staircase.

Hamilton's great contribution to optics was unification. He took a huge variety of known results and reduced them all to the same fundamental technique. In place of a system of light rays he introduced a single quantity, the "characteristic function" of the system. Any optical configuration was thereby represented by a single equation. Furthermore, this equation could be solved by a uniform method, leading to a complete depiction of the system of rays and its behavior. The method rested on a single fundamental principle: that light rays traveling through any system of mirrors, prisms, and lenses will follow the path that gets the light to its destination in the shortest time.

Fermat had already found some special cases of this principle, calling it the Principle of Least Time. The easiest example to explain it is light reflecting from a flat mirror. The left-hand figure below shows a light ray emerging from one point and bouncing off the mirror to reach a second point. One of the great early discoveries in optics was the law of reflection, which states that the two segments of the light ray make equal angles with the mirror.

Fermat came up with a neat trick: reflect the second segment of the ray, and the second point, in the mirror, as in the right-hand figure. Thanks to Euclid, the "equal angles" condition is the same as the statement that in this reflected representation, the path from the first point to the second is straight. But Euclid famously proved that a straight line is the shortest path between two points. Since the speed of light in air is constant, shortest *distance* equates to shortest *time*. "Unreflecting" the geometry to get back to the left-hand figure, the same statement holds. So the equal-angles condition is logically equivalent to the light ray taking the shortest time to get from the first point to the second, subject to its hitting the mirror along the way.

A related principle, Snell's law of refraction, tells us how light rays bend when passing from air into water, or from any medium to any other. It can be derived by a similar method, bearing in mind that light travels more slowly in water than it does in air. Hamilton went further, declaring that the same principle of minimizing time applied to all optical systems, and

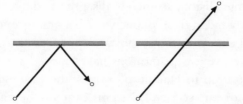

How the principle of least time leads to the law of reflection.

capturing that thought in a single mathematical object, the characteristic function.

The mathematics was impressive, but in Hamilton's hands it led to immediate experimental payoff. Hamilton noticed that his method implied the existence of "conical refraction," in which a single light ray hitting a suitable crystal would emerge as an entire cone of rays. In 1832 this prediction, which was a big surprise to everyone who worked in optics, was dramatically confirmed by Humphry Lloyd using a crystal of the mineral aragonite. Overnight, Hamilton became a household name in science.

By 1830, Hamilton was thinking about settling down and considered marrying Ellen de Vere, telling Wordsworth he "admired her mind." Again he resorted to sending her poems, and was just getting ready to propose when she told him she could never leave her home village of Curragh. He interpreted this as a tactful brush-off, and he may have been right, because a year later she married someone else and moved away.

Eventually, he married Helen Bayly, a local lass who lived near the observatory. Hamilton described her as "not at all brilliant." The honeymoon was a disaster; Hamilton worked on optics and Helen was ill. In 1834 they had a son, William Edwin. Then Helen went away for most of a year. A second son, Archibald Henry, followed in 1835, but the marriage was falling apart.

Posterity holds Hamilton's mechanico-optical analogy to have been his greatest discovery. But in his own mind, right up to his death and with increasing obsession, that honor was reserved for something very different: quaternions.

Quaternions are an algebraic structure, a close relative of complex numbers. Hamilton was convinced they held the key to the deepest regions of physics; indeed, in his final years he believed they held the key to virtually everything. History seemed to disagree, and over the next century quaternions slowly faded from public view, becoming an obscure backwater of abstract algebra with few important applications.

Very recently, however, quaternions have enjoyed a revival. While they may never measure up to Hamilton's hopes, they are increasingly recognized as an important source of significant mathematical structures. Quaternions, it turns out, are very special beasts, and they are special in just the way modern theories of physics require.

When first discovered, quaternions started a major revolution in algebra. They broke one of the important algebraic rules. Within twenty years, virtually all of the rules of algebra were routinely being broken, often bringing huge benefits, equally often leading to sterile dead ends. What the mathematicians of the mid-1850s had considered inviolable rules turned out to be merely convenient assumptions that made life simpler for algebraists but did not always match the deeper needs of mathematics itself.

In the brave new post-Galois world, algebra was no longer just about using symbols for numbers in equations. It was about the deep structure of equations—not numbers but processes, transformations, symmetries. These radical innovations changed the face of mathematics. They made it more abstract, but also more general and more powerful. And the whole area had a weird, often baffling beauty.

Until the Renaissance mathematicians of Bologna started wondering whether minus one could have a sensible square root, all of the numbers appearing in mathematics belonged to a single system. Even today, as a legacy of historical confusion about the relation of mathematics to reality, this system is known as the *real numbers.* The name is unfortunate, because it suggests that these numbers somehow belong to the fabric of the universe rather than being generated by human attempts to understand it. They are not. They are no more real than the other "number systems" invented by human imagination over the past 150 years. They do, however, bear a more direct relation to reality than most new systems. They correspond very closely to an idealized form of measurement.

A real number, essentially, is a decimal. Not as regards that specific type of notation, which is merely a convenient way to write real numbers in a form suitable for calculations, but as regards the deeper properties that decimals possess. The real numbers were born from simpler, less ambitious ancestors. First, humanity stumbled its way towards the system of "natural numbers," 0, 1, 2, 3, 4, and so on. I say "stumbled" because in the early stages, several of these numbers were not recognized as numbers at all. There was a time when ancient Greeks refused to consider 2 a number; it was too small to be typical of "numerosity." Numbers began at 3. Eventually, they allowed that 2 was as much a number as 3, 4, or 5, but then they balked at 1. After all, if someone claimed to have "a number of cows," and then you found that he owned just *one* cow, he was guilty of wild exaggeration. "Number" surely meant "plurality," which ruled out the singular.

But as notational systems developed, it became blindingly obvious that 1 was just as much a part of the computational system as its larger brethren. So it became a number—but a special, very small one. In some ways it was the most important number of all, because that's where numbers started. Adding lots of 1's together got you everything else—and for a time the notation did literally that, so that "seven" would be written as seven strokes, |||||||.

Much later, Indian mathematicians recognized that there was an even more important number that *preceded* 1. That wasn't where numbers started, after all. They started at zero—now symbolized by 0. Later still, it turned out to be useful to throw negative numbers into the mix—numbers *less than nothing*. So the negative whole numbers joined the system, and humanity invented the integers: . . ., –3, –2, –1, 0, 1, 2, 3, . . . But it didn't stop there.

The problem with whole numbers is that they fail to represent many useful quantities. A farmer trading grain, for instance, might wish to specify a quantity of wheat somewhere between 1 sack and 2 sacks. If it seemed about midway between, then it constituted 1½ sacks. Maybe it was a bit less, 1⅓, or a bit more, 1⅔. And so fractions were invented, with a variety of notations. Fractions interpolated between the whole numbers. Sufficiently complicated fractions interpolated exceedingly finely, as we have already seen with Babylonian arithmetic. Surely any quantity could be represented by a fraction.

Enter Pythagoras and his eponymous theorem. An immediate consequence is that the length of the diagonal of a unit square is a number whose square is exactly 2. That is to say, the diagonal has length equal to the square root of 2. Such a number must exist, because you can draw a square and it obviously *has* a diagonal, and the diagonal must have length. But as Hippasus realized to his sorrow, whatever the square root of 2 might be, it cannot be an exact fraction. It is irrational. So even more numbers were needed to fill invisible gaps in the system of fractions.

✳

Eventually, this process seemed to halt. The Greeks abandoned numerical schemes in favor of geometry, but in 1585, a Flemish mathematician and engineer named Simon Stevin, who lived in the town of Bruges, was appointed by William the Silent to tutor his son Maurice of Nassau. Stevin rose to become inspector of the dikes, quartermaster-general of the army, and minister of finance. These appointments, especially the last two, impressed on him the need for proper bookkeeping, and he borrowed the clerical systems of the Italians. Seeking a way of representing fractions that had the flexibility of Hindu-Arabic place notation and the fine precision of Babylonian sexagesimals, Stevin came up with a base-ten analogue of the Babylonian base-60 system: decimals.

Stevin published an essay describing his new notational system. He was sufficiently alert to marketing issues to include a statement that the ideas had been subjected to "a thorough trial by practical men who found it so useful that they had voluntarily discarded the short cuts of their own invention to adopt this new one." Further, he claimed that his decimal system "teaches how all computations that are met in business may be performed by integers alone without the aid of fractions."

Stevin's notation does not use today's decimal point, but it is directly related. Where we would write 3.1416, Stevin wrote 3 ⓪ 1 ① 4 ② 1 ③ 6 ④. The symbol ⓪ indicated a whole number, ① indicated one tenth, ② one hundredth, and so on. As people got used to the system they dispensed with ①, ②, etc., and retained only ⓪, which mutated into the decimal point.

We can't actually write the square root of two in decimals—not if we ever plan to stop. But neither can we write the fraction ⅓ in decimals. It is close to 0.33, but 0.333 is closer, and 0.3333 is closer still. An exact representation exists—to use that word in a novel way—only if we contemplate an infinitely long list of 3's. But if that's acceptable, we can in principle write down the square root of two exactly. There's no evident pattern in the digits, but by taking enough of them we can get a number whose square is as close to 2 as we please. Conceptually, if we take *all* of them, we get a number whose square is exactly 2.

With the acceptance of "infinite decimals," the real number system was complete. It could represent any number required by a businessman or mathematician to any desired accuracy. Every conceivable measurement could be stated as a decimal. If it was useful to write down negative numbers, the decimal system handled them with ease. But no other kind of number could possibly be needed. There were no gaps left to fill.

Except.

That confounded cubic formula of Cardano's seemed to be telling us something, but whatever it was, it was terribly obscure. If you started with an apparently harmless cubic—one where you *knew* a root—the formula did not give you that answer explicitly. Instead it offered a messy recipe requiring you to take cube roots of things that were even messier, and those things seemed to ask for the impossible, the square root of a negative number. The Pythagoreans had balked at the square root of two, but the square root of minus one was even more baffling.

For several hundred years, the possibility of giving sensible meaning to the square root of minus one had flitted in and out of the collective mathematical consciousness. No one had any idea whether such a number could exist. But they began to realize that it would be extremely useful if it did.

At first, such "imaginary" quantities had exactly one use: to indicate that a problem had no solutions. If you wanted to find a number whose square was minus one, the formal solution "square root of minus one" was imaginary, so no solution existed. No less a thinker than René Descartes made just this point. In 1637 he distinguished "real" numbers from "imaginary" ones, insisting that the presence of imaginaries signals the absence of solutions. Newton said the same thing. But both of these luminaries were reckoning without Bombelli, who had noticed centuries earlier that sometimes imaginaries signal the *presence* of solutions. But the signal is hard to decipher.

In 1673 the English mathematician John Wallis, who was born in Ashford, about fifteen miles from my hometown in the county of Kent, made a fantastic breakthrough. He found a simple way to represent imaginary numbers—even "complex" ones that combined real numbers with imaginaries—as points in the plane. The first step is the now-familiar concept of the real "number line," a kind of ruler extending to infinity in both directions, with 0 in the middle, the positive real numbers wandering off to the right, and the negative ones to the left.

Every real number can be located on the number line. Each successive decimal place requires a subdivision of the unit length into ten, a hundred, a thousand, etc., equal parts, but that is no problem. Numbers like $\sqrt{2}$ can be located as accurately as we wish, somewhere in between 1 and 2, a bit to the left of 1.5. The number π sits a little to the right of 3, and so on.

But where does $\sqrt{-1}$ go? There is no place for it on the real number line. It is

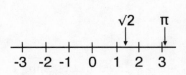

The real number line.

neither positive nor negative; it can go neither to the right nor to the left of 0.

So Wallis put it somewhere else. He introduced a second number line, to include the imaginaries—the multiples of *i*—and placed it *at right angles* to the real number line. It was literally a case of "lateral thinking."

The two number lines, real and imaginary, have to cross at 0. It is very easy to prove that if numbers make sense at all, then 0 times *i* must equal 0, so the origins of the real and imaginary lines coincide.

A complex number consists of two parts: one real, one imaginary. To locate this number in the plane, Wallis told his readers to measure off the real part along the horizontal "real" line, and then measure off the imaginary part vertically—parallel to the imaginary line.

This proposal completely solved the problem of giving meaning to imaginary and complex numbers. It was simple but decisive, a true work of genius.

It was totally ignored.

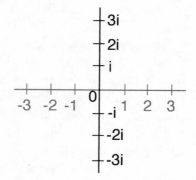

Two copies of the real number line, placed at right angles.

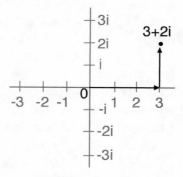

The complex plane according to Wallis.

Despite the lack of public recognition, Wallis's breakthrough must have percolated into the mathematical consciousness, because mathematicians started to employ subconscious images that related directly to Wallis's basic idea: there is no complex number line, there is a complex *plane*.

As mathematics became more versatile, mathematicians started trying to calculate ever more complicated things. In 1702, Johann Bernoulli, trying to solve a calculus problem, found that he needed to evaluate the logarithm of a complex number. By 1712, Bernoulli and Leibniz were doing battle over a core issue: what is the logarithm of a negative number? If you could solve that, you could find the logarithm of any complex number because the logarithm of a number's square root is just half the logarithm of that number. So the logarithm of i is half that of -1. But what is the logarithm of -1?

The issue at stake was simple. Leibniz believed the logarithm of -1 had to be complex. Bernoulli said it had to be real. Bernoulli based his contention on a simple piece of calculus; Leibniz objected that neither the method nor the answer made sense. In 1749, Euler resolved the controversy, coming down heavily in favor of Leibniz. Bernoulli, he pointed out, had forgotten something. His calculus calculation was of a kind that involved the addition of an "arbitrary constant." In his enthusiasm for complex calculus, Bernoulli had tacitly assumed that this constant was zero. It wasn't. It was imaginary. This omission explained the discrepancy between Bernoulli's answer and Leibniz's.

The pace of "complexification" of mathematics was heating up. More and more ideas originating in the study of real numbers were being extended to complex ones. In 1797, a Norwegian named Caspar Wessel published a method to represent complex numbers as points in a plane. Caspar came from a family of church ministers and was the sixth of fourteen children. At that time, Norway had no universities but was united with Denmark, so in 1761 he went to the University of Copenhagen. He and his brother Ole studied law, and Ole worked on the side as a surveyor, to stretch the family finances. Later, Caspar became Ole's assistant.

While working as a surveyor, Caspar invented a way to represent the geometry of the plane—especially its lines and their directions—in terms of complex numbers. Conversely, his ideas could be seen as representing complex numbers in terms of the geometry of the plane. He presented the work—his one and only research paper in mathematics—to the Royal Danish Academy in 1797.

Hardly any leading mathematicians read Danish, and the work languished unread until it was translated into French a century later. Meanwhile, the French mathematician Jean-Robert Argand independently had the same idea and published it in 1806. By 1811 it had occurred to Gauss, independently again, that complex numbers could be viewed as the points of a plane. The terms "Argand diagram," "Wessel plane," and "Gauss plane" began to circulate. Different nationalities tended to employ different phrases.

Hamilton took the final step. In 1837, almost three hundred years after Cardano's formula had suggested that "imaginary" numbers might be useful, Hamilton removed the geometric element and reduced complex numbers to pure algebra. His idea was simple; it was implicit in Wallis's proposal and in the equivalent ideas of Wessel, Argand, and Gauss. But no one had made it explicit.

Algebraically, said Hamilton, a point in the plane can be identified with a pair of real numbers, its coordinates (x, y). If you look at Wallis's diagram (or Wessel's, Argand's, or Gauss's) you can see that x is the real part of the number and y is its imaginary part. A complex number $x + iy$ is "really" just a pair of real numbers, (x, y). You can even lay down rules for adding and multiplying these pairs, and the main step is to observe that since i corresponds to the pair $(0, 1)$, then $(0, 1) \times (0, 1)$ must equal $(-1, 0)$. At this point Gauss revealed, in a letter to the Hungarian geometer Wolfgang Bolyai, that exactly the same idea had occurred to him in 1831. Once again, the fox had covered his tracks—so completely that nothing had been visible.

Problem solved. A complex number is just a pair of real numbers, manipulated according to a short list of simple rules. Since a pair of real numbers is surely just as "real" as a single real number, real and complex numbers are equally closely related to reality, and "imaginary" is misleading.

Today's view is rather different: it is that "real" is what's misleading. Both real and imaginary numbers are equally figments of human imagination.

The reaction to Hamilton's solution of a three-hundred-year-old conundrum was distinctly muted. Once mathematicians had woven the notion of complex numbers into a powerful coherent theory, fears about the existence of complex numbers became unimportant. But Hamilton's use of pairs turned out nonetheless to be very significant. Even though the issue of complex numbers was no longer a source of excitement, the idea of building new number systems from old ones took root in the mathematical consciousness.

Complex numbers, it turned out, were useful not only in algebra and basic calculus. They constituted a powerful method for solving problems about fluid flow, heat, gravity, sound, almost every area of mathematical physics. But they had one major limitation: they solved these problems in two-dimensional space, not the three that we live in. Some problems, such as the motion of the skin of a drum or the flow of a thin layer of fluid, can be reduced to two dimensions, so the news isn't all bad. But mathematicians became increasingly irritated that their complex-number methods could not be extended from a plane to space of three dimensions.

Might there be an undiscovered extension of the number system to three dimensions? Hamilton's formalization of complex numbers as *pairs* of real numbers suggested a way to approach this proposal: try to set up a number system based on *triples* (x, y, z). The problem was that no one had worked out an algebra of triples. Hamilton decided to try.

Adding triples was easy: you could take a hint from complex numbers and just add corresponding coordinates. This kind of arithmetic, known today as "vector addition," obeys very pleasant rules, and there is only one sensible way to do it.

The bugbear was multiplication. Even for the complex numbers, multiplication does not work like addition. You do not multiply two pairs of real numbers by separately multiplying their first and second components. If you do, a lot of pleasant things happen—but two fatally unpleasant ones happen as well.

The first is that there is no longer a square root of minus one.

The second is that you can multiply two nonzero numbers together and get zero. Such "divisors of zero" play merry hell with all of the usual algebraic methods, such as ways to solve equations.

For the complex numbers we can overcome this obstacle by choosing a less obvious rule for multiplication, which is what Hamilton did. But when he tried similar tricks on triples of numbers, he got a horrible shock. Try as he might, he could not avoid some fatal defect. He could get a square root for minus one, but only by introducing divisors of zero. Getting rid of divisors of zero seemed to be completely impossible, whatever else he did.

If you're thinking that this sounds a bit like attempts to solve the quintic, you're onto something. If many capable mathematicians try something and fail, it is conceivable that it may be impossible. If there is one big thing mathematics has taught us, it is that many problems do not have solutions. You can't find a fraction whose square is 2. You can't trisect an angle with straightedge and compass. You can't solve the quintic by radicals. Mathematics has limits. Maybe you can't construct a three-dimensional algebra with all the nice properties you would like it to have.

If you're serious about finding out whether that is indeed the case, a program of research opens up. First you need to specify the properties you want your three-dimensional algebra to have. Then you must analyze the consequences of those properties. Given enough information from that program, you can search for features that such an algebra must have if it does exist, and reasons why it might not exist.

At least, that's what you would do today. Hamilton's approach was not so systematic. He tacitly assumed that his algebra must have "all" reasonable properties, and suddenly realized that one of them might have to be dispensed with. More significantly, he realized that an algebra of three dimensions was not in the cards. The closest he could get was four. Quadruples, not triples.

Back to those elusive rules of algebra. When mathematicians do algebraic calculations, they rearrange symbols in systematic ways. Recall that the original Arabic name "al-jabr" means "restoration"—what nowadays we call "move the term to the other side of the equation and change its sign." Only within the last 150 years have mathematicians bothered to make explicit lists of the rules behind such manipulations, deriving other well-known rules as logical consequences. This axiomatic approach does for algebra what Euclid did for geometry, and it took mathematicians only two thousand years to get the idea.

To set the scene, we can focus on three of these rules, all related to multiplication. (Addition is similar but more straightforward; multiplication is where everything starts to go pear-shaped.) Children learning their multiplication tables eventually notice some duplication of effort. Not only does three times four make twelve: so does four times three. If you multiply two numbers together you get the same result whichever one comes first. This fact is called the *commutative law*, and in symbols it tells us that $ab = ba$ for any numbers a and b. This rule also holds in the extended system of complex numbers. You can prove this by examining Hamilton's formulas for how to multiply pairs.

A subtler law is the *associative law*, which says that when you multiply three numbers together in the same order, it makes no difference where you start. For example, suppose I want to work out $2 \times 3 \times 5$. I can start with 2×3, getting 6, and then multiply 6 by 5. Alternatively, I can start with 3×5, which is 15, and then multiply 2 by 15. Either method yields the same result, namely 30. The associative law states that this is always the case; in symbols it says that $(ab)c = a(bc)$, where the parentheses show the two ways to do the multiplication. Again, this rule holds for both real and complex numbers, and this can be proved using Hamilton's formulas.

A final, very useful rule—let me call it the *division law,* although you will find it in the textbooks as "existence of a multiplicative inverse"—states that you can always *divide* any number whatsoever by any nonzero number. There are good reasons to forbid division by zero: the main one is that it seldom makes sense.

We saw earlier that you can manufacture an algebra of triples using an "obvious" form of multiplication. This system satisfies the commutative law and the associative law. But it fails to obey the division law.

Hamilton's great inspiration, reached after hours of fruitless searching and calculation, was this: it is possible to form a new number system in which both the associative law and the division law are valid, but you have to sacrifice the commutative law. Even then, you can't do it with triples of real numbers. You have to use quadruples. There is no "sensible" three-dimensional algebra, but there is a fairly nice *four*-dimensional one. It is the only one of its kind, and it falls short of the ideal in just one respect: the commutative law fails.

Does that matter? Hamilton's biggest mental block was in thinking that the commutative law was essential. All that changed in an instant when, inspired by who knows what, he suddenly understood how to multiply quadruples. The date was 16 October 1843. Hamilton and his wife were walking along the towpath of the Royal Canal, heading for a meeting of the prestigious Royal Irish Academy in Dublin. His subconscious mind must have been churning away at the problem of three-dimensional algebra, because inspiration suddenly struck. "I then and there felt the galvanic circuit of thought *close,*" he wrote in a subsequent letter, "and the sparks which fell from it were the *fundamental equations between i, j, k; exactly such* as I have used them ever since."

So overcome was Hamilton that he immediately carved the formulas into the stonework of Broome Bridge (he called it "Brougham"). The bridge survives, but not the carving—though there is a commemorative plaque. The formulas also survive:

$$i^2 = j^2 = k^2 = ijk = -1$$

These are very pretty formulas, with a lot of symmetry. But what you are probably wondering is, where are the quadruples?

Complex numbers can be written as pairs (x, y), but they are usually written as $x + iy$ where $i = \sqrt{-1}$. In the same manner, the numbers Hamilton had in mind could be written either as quadruples (x, y, z, w) or as a combination $x + iy + jz + kw$. Hamilton's formulas use the second notation; if you are of a formal turn of mind, you may prefer to use quadruples instead.

Hamilton called his new numbers *quaternions*. He proved that they obey the associative law and—remarkably, as it later transpired—the division law. But not the commutative law. The rules for multiplying quaternions imply that $ij = k$, but $ji = -k$.

The system of quaternions contains a copy of the complex numbers, the quaternions of the form $x + iy$. Hamilton's formulas show that -1 does not have just two square roots, i and $-i$. It also has j, $-j$, k, and $-k$. In fact there are infinitely many different square roots of minus one in the quaternion system.

So along with the commutative law, we have also lost the rule that a quadratic equation has two solutions. Fortunately, by the time quaternions were invented, the focus of algebra had shifted away from the solution of equations. The advantages of quaternions greatly outweighed their defects. You just had to get used to them.

In 1845, Thomas Disney visited Hamilton and brought his daughter, William's childhood love Catherine, with him. By then she had lost her first husband and married again. The encounter reopened old wounds, and Hamilton's reliance on alcohol became more severe. He made such a complete fool of himself at a scientific dinner in Dublin that he went on the wagon and drank only water for the next two years. But when the astronomer George Airy began taunting him for his abstinence, Hamilton responded by downing alcohol in large quantities. From then on he was a chronic alcoholic.

Two uncles died, and a friend and colleague committed suicide; then Catherine started writing to him, which made his depression worse. She quickly realized that what she was doing was not proper for a respectable married woman, and made a half-hearted attempt to kill herself. She separated from her husband and went to live with her mother.

Hamilton kept writing to her, through her relatives. By 1853 she had renewed contact, sending him a small gift. Hamilton responded by going to see her, bearing a copy of his book on quaternions. Two weeks later, she was dead, and Hamilton was grief-stricken. His life became more and more disorderly; uneaten food was found mixed with his mathematical papers after his death, which occurred in 1865—attributed to gout, a common disease of heavy drinkers.

Hamilton believed quaternions to be the Holy Grail of algebra and physics—the true generalization of complex numbers to higher dimensions, and the key to geometry and physics in space. Of course, space has three dimensions, while quaternions have four, but Hamilton spotted a natural subsystem with three dimensions. These were the "imaginary" quaternions $bi + cj + dk$. Geometrically, the symbols i, j, k can be interpreted as rotations about three mutually perpendicular axes in space, although there are some subtleties: basically, you have to work in a geometry where a full circle contains 720°, not 360°. This quirk aside, you can see why Hamilton found them useful for geometry and physics.

The missing "real" quaternions behaved just like real numbers. You couldn't eliminate them altogether, because they were likely to turn up whenever you carried out algebraic calculations, even if you started with imaginary quaternions. If it had been possible to stay solely within the domain of imaginary quaternions, there would have been a sensible three-dimensional algebra, and Hamilton's quest would have succeeded. The four-dimensional system of quaternions was the next best thing, and the natural three-dimensional system embedded rather tidily inside it was just as useful as a purely three-dimensional algebra would have been.

Hamilton devoted the rest of his life to quaternions, developing their mathematics and promoting their applications to physics. A few devoted followers sung their praises. They founded a school of quaternionists, and when Hamilton died the reins were taken up by Peter Tait in Edinburgh and Benjamin Peirce at Harvard.

Others, however, disliked quaternions—partly for their artificiality, but mostly because they believed they had found something better. The most prominent of the dissenters were the Prussian Hermann Grassmann and the American Josiah Willard Gibbs, now recognized as the creators of "vector algebra." Both of them invented useful types of algebra with any number of dimensions. In their work there was no limit to four dimensions or to the three-dimensional subset of imaginary quaternions. The algebraic properties of these vector systems were not as elegant as Hamilton's quaternions. You couldn't divide one vector by another, for instance. But Grassmann and Gibbs preferred general concepts that worked, even if they lacked a few of the usual features of numbers. It may have been impossible to divide one vector by another, but who *cared?*

Hamilton went to his grave believing that quaternions were his greatest contribution to science and mathematics. For the next hundred years, hardly anyone save Tait and Peirce would have agreed, and quaternions remained an obsolete backwater of Victorian algebra. If you wanted an example of the sterility of mathematics for its own sake, quaternions were just the ticket. Even in university courses on pure mathematics, quaternions never appeared or were shown as a curiosity. According to Bell,

> Hamilton's deepest tragedy was neither alcohol nor marriage but his obstinate belief that quaternions held the key to the mathematics of the physical universe. History has shown that Hamilton tragically deceived himself when he insisted "I still must assert that this discovery appears to me to be as important for the middle of the 19th Century as the discovery of fluxions was for the close of the seventeenth." Never was a great mathematician so hopelessly wrong.

Really?

Quaternions may not have developed quite along the lines that Hamilton laid down, but their importance grows every year. They have become absolutely fundamental to mathematics, and we will see that the quaternions and their generalizations are fundamental to physics, too. Hamilton's obsession opened the door to vast tracts of modern algebra and mathematical physics.

Never was a quasi-historian so hopelessly wrong.

Hamilton may have exaggerated the applications of quaternions, and tortured them into performing tricks to which they were not really suited, but his faith in their importance is beginning to appear justified. Quaternions have developed a strange habit of turning up in the most unlikely places. One reason is that they are unique. They can be characterized by a few reasonable, relatively simple properties—a selection of the "laws of arithmetic," omitting only one important law—and they constitute the only mathematical system with that list of properties.

This statement requires unpacking.

The only number system that is familiar to most people on the planet is the real numbers. You can add, subtract, multiply, and divide real numbers, and your result is always a real number. Of course, division by zero is not tolerated, but aside from that necessary limitation, you can apply lengthy series of arithmetic operations without ever leaving the system of real numbers.

Mathematicians call such a system a *field*. There are many other fields, such as the rationals and the complex numbers, but the real field is special. It is the only field with two further properties: it is ordered, and it is complete.

"Ordered" means that the numbers occur in a linear order. The reals are strung out along a line, with negative numbers to the left and positive numbers to the right. There are other ordered fields, such as the rational numbers, but unlike the other ordered fields the reals are also complete. This extra property (whose full statement is somewhat technical) is the one that allows numbers like $\sqrt{2}$ and π to exist. Basically, the completeness property says that infinite decimals make sense.

It can be proved that the real numbers constitute the only complete ordered field. That is why they play such a central role in mathematics. They are the only context in which arithmetic, "greater than," and basic operations of calculus can be carried out.

The complex numbers extend the real numbers by throwing in a new kind of number, the square root of minus one. But the price we pay for being able to take square roots of negative numbers is the loss of order. The complex numbers are a complete system but are spread out across a plane rather than aligned in a single orderly sequence.

The plane is two-dimensional, and two is a finite integer. The complex numbers are the only field that contains the real numbers and has finite dimension—other than the real numbers themselves, with dimension one. This implies that the complex numbers, too, are unique. For many important purposes, the complex numbers are the only gadget that can do the job. Their uniqueness makes them indispensable.

The quaternions arise when we try to extend the complex numbers, increasing the dimension (while keeping it finite) and retaining as many of the laws of algebra as possible. The laws we want to keep are all the usual properties of addition and subtraction, most of the properties of multiplication, and the possibility of dividing by anything other than zero. The sacrifice this time is more serious; it is what caused Hamilton so much heartache. You have to abandon the commutative law of multiplication. You just have to accept that as a brutal fact, and move on. When you get used to it, you wonder why you ever expected the commutative law to hold in any case, and start to think it a minor miracle that it holds for the complex numbers.

Any system with this mix of properties, commutative or not, is called a *division algebra.*

The real numbers and the complex numbers are division algebras, because we don't rule out commutativity of multiplication, we just don't demand it. Every field is a division algebra. But some division algebras are not fields, and the first to be discovered was the quaternions. In 1898, Adolf Hurwitz proved that the system of quaternions is also unique. The quaternions are the *only* finite-dimensional division algebra that contains the real numbers and is not equal either to the real numbers or to the complex numbers.

There is a curious pattern here. The dimensions of the reals, complexes, and quaternions are 1, 2, and 4. This looks suspiciously like the start of a sequence, the powers of 2. A natural continuation would be 8, 16, 32, and so on.

Are there interesting algebraic systems with those dimensions?

Yes and no. But you'll have to wait to see why, because the story of symmetry now enters a new phase: connections with differential equations, the most widely used way to model the physical world, and the language in which most of the physicists' laws of nature are couched.

Again, the deepest aspects of the theory boil down to symmetry, but with a new twist. Now the symmetry groups are not finite, but "continuous." Mathematics was about to be enriched by one of the most influential programs of research ever conducted.

10

THE WOULD-BE SOLDIER
AND THE WEAKLY BOOKWORM

arius Sophus Lie studied science only because his poor eyesight dis-
qualified him from any military profession. When Sophus, as he
came to be called, graduated from the University of Christiania in
1865, he had taken a few mathematics courses, including one on Galois theory given by the Norwegian Ludwig Sylow, but he showed no special talent in the subject. For a while he dithered—he knew that he wanted
an academic career but was unsure whether it should be in botany, zool-
ogy, or perhaps astronomy.

The library records at the university show him taking out more and
more books on mathematical topics. In 1867, in the middle of the night,
he was struck by a vision of his life's work. His friend Ernst Motzfeldt was
astonished to be woken from sleep by an excited Lie, who was shouting "I
have found it, it is quite simple!"

What he had found was a new way to think about geometry.

Lie began to study the works of the great geometers, such as the German Julius Plücker and the Frenchman Jean-Victor Poncelet. From
Plücker he got the idea of geometries whose underlying elements are not
Euclid's familiar points but other objects—lines, planes, circles. He published a paper outlining his big idea in 1869, at his own expense. Like Galois and Abel before him, he discovered that his ideas were too
revolutionary for the old guard, and the regular journals did not wish to
publish his researches. But Ernst refused to let his friend become discour-
aged, and kept him working on his geometry. Eventually, one of Lie's

papers was published in a prestigious journal and was favorably received. It gained Lie a scholarship. Now he had the money to travel, visit leading mathematicians, and discuss his ideas with them. He went to the hotbeds of Prussian and German mathematics, Göttingen and Berlin, and talked to the algebraists Leopold Kronecker and Ernst Kummer and the analyst Karl Weierstrass. He was impressed by Kummer's way of doing mathematics, less so by Weierstrass's.

The most significant meeting, however, was in Berlin with Felix Klein—who, it happened, had been a student of Plücker, whom Lie greatly admired and wished to emulate. Lie and Klein had very similar mathematical backgrounds, but their tastes differed considerably. Klein, basically an algebraist with geometrical leanings, enjoyed working on special problems with inner beauty; Lie was an analyst who liked the broad sweep of general theories. Ironically, it was Lie's general theories that gave mathematics some of its most important special structures, which were and still are extraordinarily beautiful, extraordinarily deep, and mostly algebraic. These structures might not have been discovered at all were it not for Lie's push to generality. If you try to understand all possible mathematical objects of a certain kind, and succeed, you will inevitably find many that have unusual features.

In 1870, Lie and Klein met again in Paris. And there, Jordan converted Lie to the cause of group theory. There was a growing realization that geometry and group theory were two sides of the same coin, but it took a long time for the idea to become fully formed. Lie and Klein did some joint work, trying to make the connection between groups and geometry more explicit. Eventually, Klein crystallized the thought in his "Erlangen Program" of 1872, according to which geometry and group theory are identical.

In modern language, the idea sounds so simple that it should have been obvious all along. The group that corresponds to any given geometry is the *symmetry group* of that geometry. Conversely, the geometry corresponding to any group is whatever object it is the symmetry group *of.* That is, the geometry is defined by those things that are invariant under the group.

For example, the symmetries of Euclidean geometry are those transformations of the plane that preserve lengths, angles, lines, and circles. This is the group of all rigid motions of the plane. Conversely, anything that is invariant under rigid motions naturally falls within the purview of Euclid-

ean geometry. Non-Euclidean geometry simply employs different transformation groups.

Why, then, bother converting geometry into group theory? Because it
gives you two different ways to think about geometry, and two different
ways to think about groups. Sometimes one way is easier to understand,
sometimes the other. Two points of view are better than one.

Relations between France and Prussia were deteriorating fast. Emperor
Napoleon III thought he could shore up his declining popularity by starting a war with Prussia. Bismarck sent the French a stinging telegram, and
the Franco-Prussian War was declared on 19 July 1870. Klein, a Prussian
in Paris, deemed it prudent to head back to Berlin.

Lie, however, was Norwegian and was greatly enjoying his visit, so he
decided to stay in Paris. But he changed his mind when he realized that
France was losing the war and the German army was advancing on Metz.
Although he was a citizen of a neutral country, it was not safe to remain in
a potential war zone.

Lie decided to go on a hiking trip, heading for Italy. He did not get far;
the French authorities caught him at Fontainebleau, about 25 miles south-
east of Paris, carrying a number of documents covered in incomprehensible symbols. Since these were evidently in code, Lie was obviously spying
for the Germans, and he was placed under arrest. It took the intervention
of a leading French mathematician, Gaston Darboux, to convince the authorities that the writings were mathematics. Lie was let out of prison, the
French army surrendered, the Germans began a blockade of Paris, and
Lie once more headed for Italy—this time successfully. From there he returned to Norway. Along the way he dropped in on Klein, who had remained safe in Berlin.

Lie received his doctorate in 1872. The Norwegian academic world was
so impressed by his work that the University of Christiania created a position especially for him in the same year. With his former teacher Ludwig
Sylow, he took on the task of editing Abel's collected works. In 1874 he
married Anna Birch, eventually fathering three children.

By now, Lie had focused on a particular topic that he felt was ripe for
development. There are many kinds of equations in mathematics, but two
types are especially important. The first is algebraic equations, of the kind

studied so effectively by Abel and Galois. The second is differential equations, introduced by Newton in his work on the laws of nature. Such equations involve concepts from calculus, and instead of dealing directly with some physical quantity, they describe how that quantity changes as time passes. More precisely, they specify the *rate of change* of the quantity. For example, Newton's most important law of motion states that the acceleration experienced by a body is proportional to the total force acting on it. Acceleration is the rate of change of velocity. Instead of telling us directly what the body's velocity is, the law tells us the rate of change of velocity. Similarly, another equation that Newton developed, to explain how the temperature of an object changes when it cools, states that *the rate of change* of temperature is proportional to the difference between the temperature of the object and the temperature of its surroundings.

Most of the important equations in physics—those that concern the flow of a fluid, the action of gravity, the motion of the planets, the transfer of heat, the movement of waves, the action of magnetism, and the propagation of light and sound—are differential equations. As Newton first realized, nature's patterns generally become simpler and easier to spot if we look at the rates of change of the quantities that we want to observe, not at those quantities themselves.

Lie posed himself a momentous question. Is there a theory of differential equations analogous to Galois's theory of algebraic ones? Is there a way to decide when a differential equation can be solved by specified methods?

The key, once more, was symmetry. Lie now realized that some of his results in geometry could be reinterpreted in terms of differential equations. Given one solution of a particular differential equation, Lie could apply a transformation (from a particular group) and prove that the result was also a solution. From one solution you could get many, all connected by the group. In other words, the group consisted of symmetries of the differential equation.

It was a broad hint that something beautiful awaited discovery. Consider what Galois's application of symmetries had done for algebraic equations. Now imagine doing the same thing for the far more important class of differential equations!

❊

The groups studied by Galois are all finite. That is, the number of transformations in the group is a whole number. The group of all permutations on the five roots of a quintic, for example, has 120 elements. Many sensible groups are infinite, however, including symmetry groups of differential equations.

One common infinite group is the symmetry group of a circle, which contains transformations that rotate the circle through any angle whatsoever. Since there are infinitely many possible angles, the rotation group of the circle is infinite. The symbol for this group is SO(2). Here "O" stands for "orthogonal," meaning that the transformations are rigid motions of the plane, and "S" means "special"—rotations do not flip the plane over.

Circles also have infinitely many axes of reflectional symmetry. If you reflect a circle in any diameter, you get the same circle. Adding in the reflections leads to a bigger group, O(2).

The groups SO(2) and O(2) are infinite, but it is a tame type of infinity. The different rotations can all be determined by specifying a single number—the relevant angle. When two rotations are composed, you just add the corresponding angles. Lie called this kind of behavior "continuous," and in his terminology, SO(2) was therefore a continuous group. And because only one number is needed to specify an angle, SO(2) is one-dimensional. The same goes for O(2), because all we need is a way to distinguish reflections from rotations, and this is a matter of a plus or minus sign in the algebra.

The group SO(2) is the simplest example of a *Lie group,* which has two types of structure at the same time: it is a group and also a manifold—a multidimensional space. For SO(2), the manifold is a circle, and the group operation combines two points on the circle by adding the corresponding angles.

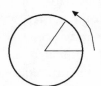 

The circle has infinitely many rotational symmetries (left) and infinitely many reflectional symmetries (right).

Lie discovered a beautiful feature of Lie groups: the group structure can be "linearized." That is, the underlying curved manifold can be replaced by a flat Euclidean space. This space is the tangent space to the manifold. Here's how it looks for SO(2):

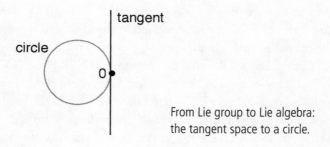

From Lie group to Lie algebra: the tangent space to a circle.

The group structure, when linearized in this fashion, gives the tangent space an algebraic structure of its own, which is a kind of "infinitesimal" version of the group structure, describing how transformations very close to the identity behave. It is called the *Lie algebra* of that group. It has the same dimension as the group, but its geometry is much simpler, being flat.

There is a price to pay for this simplicity, of course: the Lie algebra captures most important properties of the corresponding group, but some fine detail gets lost. And those properties that are captured undergo subtle changes. Nonetheless, you can learn a lot about a Lie group by passing to its Lie algebra, and most questions are more easily answered in the Lie algebra setting.

It turns out—and this was one of Lie's great insights—that the natural algebraic operation on the Lie algebra is not the product AB, but the difference $AB - BA$, which is called the *commutator*. For groups like SO(2), where $AB = BA$, the commutator is zero. But in a group like SO(3), the rotation group in three dimensions, $AB - BA$ is nonzero unless the axes of rotation of A and B are either the same or at right angles. So the geometry of the group shows up in the behavior of commutators.

Lie's dream of a "Galois theory" of differential equations was eventually realized with the creation of a theory of "differential fields" in the early 1900s. But the theory of Lie groups turned out to be far more important, and more widely applicable, than Lie expected. Instead of being a tool to determine whether a differential equation can be solved in specific ways, the theory of Lie groups and Lie algebras has pervaded almost every branch of mathematics. "Lie theory" escaped its creator and became greater than he ever imagined.

In hindsight, the reason is symmetry. Symmetry is deeply involved in every area of mathematics, and it underlies most of the basic ideas of mathematical physics. Symmetries express underlying regularities of the world, and those are what drive physics. Continuous symmetries such as rotations are closely related to the nature of space, time, and matter; they imply various conservation laws, such as the law of conservation of energy, which states that a closed system can neither gain nor lose energy. This connection was worked out by Emmy Noether, a student of Hilbert.

The next step, of course, is to understand the possible Lie groups, just as Galois and his successors sorted out many properties of finite groups. Here a second mathematician joined in the hunt.

Anna Catharina was worried about her son.

Her doctor had told her that young Wilhelm was "quite weakly and besides very awkward" and "always excited, but a completely impractical bookworm." Wilhelm's health improved as he grew, but his bookworm tendencies did not. Just before his 39th birthday, he would publish a piece of mathematical research that has been described, with justification, as "the greatest mathematical paper of all time." Such designations are of course subjective, but Wilhelm's paper would certainly be high on anyone's list.

Wilhelm Karl Joseph Killing was the son of Josef Killing and Anna Catharina Kortenbach. He had one brother, Karl, and one sister, Hedwig. Josef was a legal clerk, and Anna was a pharmacist's daughter. They were married in Burbach, on the eastern side of central Germany, and soon afterward moved to Medebach when Josef became the mayor there. Then he was made mayor of Winterberg, and after that mayor of Rüthen.

The family was quite well off and could afford a private tutor to prepare Wilhelm for the gymnasium, which in his case was in Brilon, 50 miles west of Dortmund. At school he liked classics—Latin, Hebrew, Greek. A teacher named Harnischmacher introduced him to mathematics; Wilhelm turned out to be very good at geometry, and resolved to become a mathematician. He attended what is now the Westphalian Wilhelm University of Münster, but was then merely a Royal Academy. The academy did not teach advanced mathematics, so Killing taught himself. He read Plücker's geometrical work and tried to derive some new theorems of his own. He also read Gauss's *Disquisitiones Arithmeticae.*

After two years at the Royal Academy he moved to Berlin, where the quality of mathematical teaching was much superior, and came under the influence of Weierstrass, Kummer, and Hermann von Helmholtz, a mathematical physicist who clarified the link between conservation of energy and symmetry. Killing wrote a PhD thesis on the geometry of surfaces, based on some ideas of Weierstrass, and took a job as a teacher of mathematics and physics, with a sideline in Greek and Latin.

In 1875, he married a music lecturer's daughter, Anna Commer. Their first two children, both sons, died in infancy; the next two, daughters named Maria and Anka, thrived. Later, Killing fathered two more sons.

By 1878, he had gone back to his old school, but now as a teacher. He had a heavy workload, about 36 contact hours per week, but somehow he found the time to continue his mathematical research—the greats always do. He published a series of important papers in top journals.

In 1882, Weierstrass secured Killing a professorship at the Lyceum Hosianum in Braunsberg, where he spent the next ten years. Braunsberg had no strong mathematical tradition and offered no colleagues with whom to discuss research, but Killing seems not to have needed such stimulation. For it was there that he made one of the most important discoveries in the whole of mathematics. It left him rather disappointed.

What he had hoped to achieve was hugely ambitious: a description of *all possible* Lie groups. The Lyceum did not buy the journals in which Lie published, and Killing had very little idea of Lie's work, but he independently discovered the role of Lie algebras in 1884. So Killing knew that each Lie group was associated with a Lie algebra, and he quickly recognized that Lie algebras would probably be more tractable than Lie groups, so his problem reduced to the classification of all possible Lie algebras.

This problem turns out to be desperately hard—we now know that it probably has no sensible answer, in the sense that no simple construction can produce all Lie algebras by a uniform and transparent procedure. So Killing was forced to settle for something far less ambitious: to describe the basic building blocks from which all Lie algebras can be assembled. This is a bit like wanting to describe all possible architectural styles but having to settle for a list of all possible shapes and sizes of brick.

These basic building blocks are known as *simple* Lie algebras. They are distinguished by a very similar property to Galois's idea of a simple group, one with no normal subgroups except trivial ones. In fact, a simple Lie group has a simple Lie algebra, and the converse is very nearly true as well. Amazingly, Killing succeeded in listing all possible simple Lie algebras— mathematicians call such a theorem a "classification."

In Killing's eyes, that classification was a very limited version of something far more general, and he was frustrated by several restrictive assumptions he had been forced to make in order to get anywhere. He was particularly irked by the need to assume simplicity, which forced him to switch to Lie algebras over the complex numbers rather than the reals. The former are better behaved but less directly related to the geometrical problems that fascinated Killing. Because of these self-imposed limitations, he did not consider his work worth publishing.

He did manage to make contact with Lie; not very fruitfully, as it turned out. First he wrote to Klein, who put him in touch with Lie's assistant Friedrich Engel, then working in Christiania. Killing and Engel hit it off immediately, and Engel became a staunch supporter of Killing's work, helped him get over some tricky points, and encouraged him to push the ideas further. Without Engel, Killing might have given up.

At first, Killing thought he knew the complete list of simple Lie algebras, and that these were the Lie algebras so(n) and su(n) associated with two infinite families of Lie groups: the special orthogonal groups SO(n), consisting of all rotations in n-space, and their analogues SU(n) in complex n-space, the special unitary groups. The historian Thomas Hawkins imagined "the amazement with which Engel read Killing's letter with its bold conjectures. Here was an obscure professor at a Lyceum dedicated to the training of clergymen in the far-away reaches of East Prussia, discoursing with authority and conjecturing profound theorems on Lie's theory of transformation groups."

In the summer of 1886, Killing visited Lie and Engel at Leipzig, where both now worked. Unfortunately, there was some friction between Lie and Killing; Lie never really appreciated Killing's work and generally tried to play down its significance.

Killing quickly discovered that his original conjecture about simple Lie algebras was wrong, for he discovered a new one, whose corresponding Lie group is now known as G_2. It had 14 dimensions, and unlike the special linear and orthogonal Lie algebras, it did not seem to belong to an infinite family. It was a lone exception.

If this was strange, the final classification, which Killing completed in the winter of 1887, was stranger. To the two infinite families Killing added a third, the Lie algebras $sp(2n)$ of what are now known as the symplectic groups $Sp(2n)$. (Nowadays, we split the orthogonal groups into two different subfamilies, those acting on spaces of even dimension and those acting on spaces of odd dimension, yielding four families. There are reasons for doing this.) And now the exception G_2 had acquired five companions: two of dimension 56, and a short family that petered out, with dimensions 78, 133, and 248.

Killing's classification proceeded by a lengthy algebraic argument, which reduced the entire question to a beautiful problem in geometry. From a hypothetical simple Lie algebra he conjured up a configuration of points in a multidimensional space, known today as a *root system*. For exactly three of the simple Lie algebras, the root system lives in a space of two dimensions. These root systems look like this:

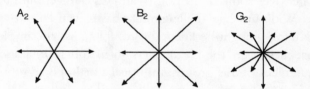

The root systems in two dimensions.

These patterns have a great deal of symmetry. In fact, they are reminiscent of the patterns you see in a kaleidoscope, where two mirrors set at an angle create multiple reflections. The similarity is no coincidence, because root systems have wonderful, elegant symmetry groups. Now known as Weyl groups (unfair, since they were invented by Killing), they are multidimensional analogues of the patterns formed by reflecting objects in a kaleidoscope.

The underlying structure of Killing's proof is that the search for all possible simple Lie algebras can be carried out by breaking the algebras into nice pieces, analogous to structures found in $su(n)$. The classification then reduces to the geometry of those pieces, using their wonderful symmetries. Having sorted out the geometry of those pieces, you can now bootstrap your results back to the problem you really wanted to solve: finding the possible simple Lie algebras.

As Killing put it, "The roots of a simple system correspond to a simple group. Conversely, the roots of a simple group can be regarded as determined by a simple system. In this way one obtains the simple groups. For each l there are four structures, supplemented for $l = 2, 4, 6, 7, 8$ by exceptional simple groups."

Here "group" was a shortened form of "infinitesimal group," which we now call a Lie algebra, and l is the dimension of the root system.

The four structures that Killing refers to are the Lie algebras $su(n)$, $so(2n)$, $so(2n+1)$, and $sp(2n)$ corresponding to families of groups $SU(n)$, $SO(2n)$, $SO(2n+1)$, and $Sp(2n)$: the unitary groups, the orthogonal groups in spaces of even dimension, the orthogonal groups in spaces of odd dimension, and the symplectic groups in spaces of even dimension. The symplectic groups are the symmetries of the position–momentum variables introduced by Hamilton in his formulation of mechanics, and the number of dimensions there is always even because the variables come in position–momentum pairs. Aside from these four families, Killing claimed that exactly six other simple Lie algebras exist.

He was nearly right. In 1894, the French geometer Élie Cartan noticed that Killing's two 56-dimensional algebras are really the same algebra viewed in two different ways. That means that there are only five exceptional simple Lie algebras, corresponding to five exceptional simple Lie groups: Killing's old friend G_2, and four others now called F_4, E_6, E_7, and E_8.

This is an exceedingly curious answer. The infinite families are reasonable enough; they are all related to various natural types of geometry in any number of dimensions. But the five exceptional Lie groups seem unrelated to anything geometric, and their dimensions are bizarre. Why are spaces of dimensions 14, 56, 78, 133, and 248 special? What is so unusual about those numbers?

It's a bit like wanting to list all possible shapes for a brick, and finding an answer something like this:

- Oblong blocks of size 1, 2, 3, 4, . . .
- Cubes of size 1, 2, 3, 4, . . .
- Slabs of size 1, 2, 3, 4, . . .
- Pyramids of size 1, 2, 3, 4, . . .

Which would be very neat and tidy, except that the list continues:

- A tetrahedron of size 14.
- An octahedron of size 52.
- A dodecahedron of size 78.
- A dodecahedron of size 133.
- A dodecahedron of size 248.

And that's it, there's nothing else.

Why do bricks with these strange shapes and sizes exist? What are they *for?*

It seemed completely mad.

It seemed so mad, in fact, that Killing was rather upset that the exceptional groups existed, and for a time he hoped they were a mistake that he could eradicate. They spoiled the elegance of his classification. But they were *there,* and we are finally beginning to understand *why* they are there. In many ways, the five exceptional Lie groups now look much more interesting than the four infinite families. They seem to be important in particle physics, as we will see; they are definitely important in mathematics. And they have a secret unity, not yet fully uncovered, relating them all to Hamilton's quaternions and an even more curious generalization, the octonions. Of which more, in due course.

It's a wonderful series of ideas, and Killing had all of them. To be sure, his work included a few mistakes—some proofs that didn't quite work. But the mistakes were all repaired long ago.

✳

That is how the greatest mathematical paper of all time went. What did Killing's contemporaries think of it?

Not a lot. It didn't help that Lie poured derision on Killing's magnum opus. He had fallen out with Killing for unknown reasons, and as far as he was concerned, Killing would never do anything important. Worse, of course, this was a theorem that Lie himself would have dearly loved to prove. Having been beaten to the punch, he resorted to the age-old technique of sour grapes. Anything in the area not done by Lie, said Lie, was rubbish. Though he wasn't quite that blatant.

It helped even less that Killing underestimated the value of his own theorem. To him it was a pale shadow of something far more important, which he had failed to achieve: classifying all Lie groups. Killing was a modest man, and Lie did his best to make him more so.

In any case, Killing was ahead of his time. Very few mathematicians saw how important Lie theory was going to become. To most, it was a rather technical branch of geometry associated with differential equations.

Finally, Killing was a staunch Catholic with a strong sense of duty and humility. He took St. Francis of Assisi as his model, and at the age of 39 he and his wife entered the Third Order of the Franciscans. He seems to have been a thoroughly decent man who worked tirelessly on behalf of his students. He was a conservative and a patriot, greatly saddened by the extreme social dissolution of Germany after World War I. His feelings were made worse by the deaths of his two sons in 1910 and 1918.

The true worth of Killing's researches became apparent in 1894, when Élie Cartan rederived the whole theory in his PhD thesis, and took it a big step further by classifying not just the simple Lie algebras but their representations in terms of matrices. Cartan was scrupulous in giving credit to Killing for nearly all of the ideas; he just tidied everything up, plugged a few gaps (some serious), and modernized the terminology. But a myth quickly grew up to the effect that Killing's work was riddled with holes and the real credit should go to Cartan. Mathematicians are seldom good historians, and they tend to cite work that they know rather than the earlier work that led up to it. So Cartan's name became attached to many of Killing's ideas.

Anyone who reads Killing's papers quickly discovers that the myth is just that. The ideas are clear and well formed, the proofs are perhaps old-fashioned but nearly all correct. Most importantly, the overall sweep of the ideas is beautifully chosen to produce the desired result. It is mathematics of the highest order, and it is not anyone else's.

Unfortunately, hardly anyone read Killing's papers. They read Cartan and ignored the credit he gave to Killing. But eventually, Killing's work began to achieve proper recognition. In 1900 he won the Lobachevsky Prize of the Kazan Physico-Mathematical Society. This was the second time the prize had been awarded: the first one went to Lie.

Killing died in 1923. Even today, his name is not as well known as it deserves to be. He was one of the greatest mathematicians who ever lived. His legacy, at least, is immortal.

11

THE CLERK FROM THE PATENT OFFICE

B y the beginning of the twentieth century, groups were starting to show up in fundamental physics, a field they would transform just as radically as they had transformed mathematics.

In the golden year of 1905, the man who would become the most iconic scientist of his time published three papers, each of which revolutionized a separate branch of physics. He was not at that time a professional scientist. He had studied at university but had not been able to obtain a teaching position and was working as a clerical official in the patent office in Bern, Switzerland. His name, of course, was Albert Einstein.

If any one person can symbolize modern physics, it is Einstein. To many, he also symbolizes mathematical genius, but in fact he was merely a competent mathematician, not a creative one on the level of Galois or Killing. Einstein's creativity lay not in producing new mathematics but in an extraordinarily rigorous intuition about the physical world, which he was able to express through remarkable uses of existing mathematics. Einstein also had a flair for the right philosophical standpoint. He drew radical theories from the simplest of principles and was guided by a sense of elegance rather than a wide knowledge of experimental facts. The important observations, he believed, could always be distilled into a few key principles. The gateway to truth was beauty.

Acres of print and many lifetimes of scholarly study have been devoted to Einstein's life and works. A single chapter cannot hope to compete in either completeness or erudition. But he is a key figure in the history of symmetry: it was Einstein, above all others, who set in motion the web of events that turned the mathematics of symmetry into fundamental physics. I don't think Einstein saw it that way: to him, the mathematics was a servant of physics—often a rather disobedient one. Only later, following the trail that Einstein had blazed and tidying up the tangled, broken vegetation that his pioneering efforts had strewn across the path, did another generation uncover the elegant and deep mathematical concepts upon which his work was based.

So we must retell the main outlines of the astonishing rise to fame of this minor patent clerk—technical expert third class, to be precise, and on a trial basis at that. Since he is but one part of our story, I will select only the relevant events. If you who want a more comprehensive, unbiased assessment of Einstein's career, you should read Abraham Pais's *Subtle Is the Lord*.

Subtle, yes—but not, as Einstein once remarked, malicious.

Einstein, who had little interest in religion, devoted his life to the principle that the universe is comprehensible and that it runs along mathematical lines. Many of his most famous sayings invoke the deity, but as a symbol of the orderliness of the universe, not as a supernatural being with a personal interest in human affairs. He worshipped no god and practiced no religious rituals.

Einstein is generally seen as the natural successor to Newton. Earlier scientists had made additions to Newton's "system of the world," as his *Mathematical Principles of Natural Philosophy* was subtitled, but Einstein was the first to make significant changes to that vision. The most important of the earlier theorists was James Clerk Maxwell, whose equations for electromagnetism brought magnetic and electric phenomena, especially light, within the Newtonian purview. Einstein went much further, making major changes. Ironically, the changes that led to a revised theory of gravity came about as consequences of the Maxwellian theory of electromagnetic waves—light and its relatives. Even more ironically, a fundamental feature of that theory, the wave nature of light, played a key role, yet Newton denied that light could be a wave. To cap it all, one of the most elegant experiments now used to demonstrate that light *is* a wave was first carried out by Newton.

Scientific interest in light goes back at least to Aristotle, who, though really a philosopher, asked the kind of question that scientists would find natural. *How do we see?* Aristotle suggested that when we look at some object, that object affects the medium between itself and the onlooking eye.

(We now call this medium "air.") The eye then detects this change in the medium, and the result is the sensation of sight.

In medieval times this explanation was reversed. It was thought that our eyes emitted some kind of ray, which illuminated whatever we looked at. Instead of the object transmitting signals to the eye, the eye left eye-tracks all over the object.

Eventually, it was understood that we see objects by means of reflected light, and that in daily life the main source of light is the Sun. Experiments showed that light travels in straight lines, forming "rays." Reflection occurs when a ray bounces off a suitable surface. So the Sun sends light rays to everything that is not shadowed by something else, the rays bounce all over the place, some enter an observer's eye, the eye receives a signal from that direction, the brain processes the incoming information from the eye, and we see whatever object the ray bounced off.

The main question was, what is light? Light does a number of puzzling things. Not only does it reflect; it can also refract—change direction abruptly at the interface between two different media, such as air and water. This is why a stick poked into a pond looks bent, and also why lenses work.

Even more puzzling is the phenomenon of diffraction. In 1664, the scientist and polymath Robert Hooke, whose career repeatedly clashed with Newton's, discovered that if he placed a lens on top of a flat mirror and then looked through the lens, he saw tiny concentric colored rings. These rings are now known as "Newton's rings" because Newton was the first person to analyze their formation. Today we consider this experiment a clear demonstration that light is a wave: the rings are interference fringes, where waves do or do not cancel each other out when they overlap. But Newton didn't believe light was a wave. Because light traveled in straight lines, he believed it had to be a stream of particles. According to his *Opticks,* completed in 1705, "Light is composed of tiny particles, or corpuscles, emitted by luminous bodies." The particle theory could explain reflection very simply: the particles bounced when they hit a (reflecting) surface. It encountered difficulties explaining refraction, and pretty much fell apart when it came to diffraction.

Thinking about what could cause light rays to bend, Newton decided that the medium, not light, must be the root cause. This led him to suggest the existence of some "aethereal medium" which transmitted vibrations *faster than light.* He convinced himself that radiant heat was evidence in favor of these vibrations, because he had established that heat radiation could traverse a vacuum. Something in the vacuum must be carrying the heat and causing refraction and diffraction. In Newton's words:

> Is not the Heat of the warm Room convey'd through the Vacuum by the Vibrations of a much subtiler Medium than Air, which after the Air was drawn out remained in the Vacuum? And is not this Medium the same with that Medium by which Light is refracted and reflected, and by whose Vibrations Light communicates Heat to Bodies, and is put into Fits of easy Reflexion and easy Transmission?

When I read these words I cannot help thinking of my friend Terry Pratchett, whose series of fantasy novels set on "Discworld" satirize our own world, and whose assorted wizards, witches, trolls, dwarves, and people poke fun at human foibles. Light on Discworld travels at roughly the speed of sound, which is why the light of dawn can be seen approaching across the fields. A necessary counterpart to light is *dark*—on Discworld almost everything is reified—and dark evidently travels faster than light because it has to get out of light's way. It all makes excellent sense, even in our world, aside from the disappointing fact that none of it is true.

Newton's theory of light suffers from the same defect. Newton wasn't being stupid: his theory seemed to answer a number of important questions. Unfortunately, these answers were based on a fundamental misunderstanding: he thought radiant heat and light were two different things. He believed that when light hits a surface, it excites heat vibrations. These were variants of the same vibrations that he thought caused light to refract and diffract.

Thus was born the concept of the "luminiferous aether," which proved remarkably persistent. Indeed, when it later turned out that light is a wave, the aether provided just the right medium for it to be a wave *in*. (We now think that light is neither wave nor particle exclusively but a bit of both—a wavicle. But I'm getting ahead of myself.)

What, though, was the aether? Newton is perfectly frank: "I do not know what this Aether is." He argued that if the aether is also composed of particles, then they must be much smaller and lighter than particles of air or even of light—essentially for the Discworldly reason that they have to be able to get out of light's way. "The exceeding smallness of its Particles," Newton says of the aether, "may contribute to the greatness of the force by which those Particles may recede from one another, and thereby make that Medium exceedingly more rare and elastick than Air, and by consequence exceedingly less able to resist the motions of Projectiles, and exceedingly more able to press upon gross Bodies, by endeavoring to expand itself."

Earlier, in his 1678 *Treatise on Light,* the Dutch physicist Christiaan Huygens had proposed a different theory: light is a wave. This theory neatly explains reflection, refraction, and diffraction—similar effects can be seen, for instance, in water waves. The aether was to light as water was to waves on the ocean—the thing that moved when the wave passed. But now Newton disagreed. The debate got very confused, because both scientists were making incorrect assumptions about the nature of the alleged waves.

Everything changed when Maxwell got in on the act. And he stood on the shoulders of another giant.

Electric heating, lighting, radio, television, food processors, microwave ovens, refrigerators, vacuum cleaners, and endless items of industrial machinery all derive from the insights of one man, Michael Faraday. Faraday was born in Newington Butts, London (now the Elephant and Castle), in 1791. He was a blacksmith's son who rose to scientific eminence in the Victorian era. His father belonged to the Sandemanians, a minority Christian sect.

Faraday became an apprentice bookbinder in 1805 and began performing scientific experiments, especially in chemistry. His interest in science grew significantly when, in 1810, he became a member of the City Philosophical Society, a group of young people who met to talk science. In 1812, he was given tickets to hear the final lectures of Sir Humphry Davy, Britain's leading chemist, at the Royal Institution. Soon thereafter, he asked Davy for a job; he was given an interview, but no position was available. But after Davy's chemical assistant was soon fired for starting a fight, Faraday got his job.

From 1813 to 1815 Faraday toured Europe with Davy and his wife. Napoleon had given Davy a passport, which included a valet, so Faraday accepted that position. He was annoyed to find that Davy's wife, Jane, took the title literally and expected him to act as her servant. In 1821, events took a more favorable turn: he was promoted, and he married

Sarah Barnard, the daughter of a prominent Sandemanian. Better still, his research into electricity and magnetism was starting to take off. Following previous research of the Danish scientist Hans Ørsted, Faraday discovered that electricity flowing through a coil near a magnet produces a force. This is the basic principle underlying the electric motor.

His research interests then became swamped under administrative and teaching duties, though these had a very favorable impact. In 1826, he started a series of evening discourses on science and also initiated the Christmas lectures for young people, both of which are still running. Today the Christmas lectures are broadcast on television, one of the gadgets that Faraday's discoveries eventually made possible. In 1831, back at his experiments, he discovered electromagnetic induction. This was the discovery that changed the industrial face of the nineteenth century, because it led to electrical transformers and generators. The experiments convinced him that electricity must be some kind of force acting between material particles, and not a fluid as generally thought.

Eminence in science typically leads to the honor of an administrative post, which promptly kills off the scientific activities that are being recognized. Faraday was made scientific adviser to Trinity House, whose mission is to keep the British seaways safe for shipping. He invented a new, more efficient kind of oil-burning lamp, which produced a brighter light. By 1840, he had become an elder of the Sandemanian sect, but his health was starting to worsen. In 1858 he was given free lodgings in a "grace and favor" house at Hampton Court, the former palace of King Henry VIII. He died in 1867 and was buried in Highgate Cemetery.

Faraday's inventions revolutionized the Victorian world, but (perhaps because of his early lack of education) he was weak on theory, and his explanations of how his inventions worked were based on curious mechanical analogies. In 1831, the year Faraday discovered how to turn magnetism into electricity, a Scottish lawyer was presented with a son—his only child, as it turned out. The lawyer was more interested in managing his land holdings, but he took considerable interest in the education of young "Jamesie," more formally known as James Clerk Maxwell.

Jamesie was bright and fascinated by machines. "How it doos?" was his standard question: How does it do that? Another was "What's the go of that?" His father, who had similar fascinations, did his best to explain.

And if the father failed to go far enough, Jamesie would ask a supplementary question: "What's the *particular* go of that?"

James's mother died of cancer when the child was nine; the loss brought father and son closer together. The boy was sent to the Edinburgh Academy, which specialized in the classics and wanted its pupils to be neat and tidy, proficient in the standard subjects, and totally lacking in original thought because that got in the way of orderly teaching. Jamesie wasn't quite what the schoolteachers wanted, and it did not help that his father, obsessed with cleanliness, had designed special clothes and shoes for the boy, including a frilly tunic bedecked with lace. The other kids nicknamed James "Dafty." But James was stubborn and earned their respect, though he still baffled them.

The school did one good thing for James: it gave him an interest in mathematics. A letter to his father talks of making "a tetra hedron, a dodeca hedron, and two more hedrons that I don't know the wright names for." (Presumably these were the octa and icosa.) By the age of 14 he had won a prize for independently inventing a class of mathematical curves known as Cartesian ovals, after its original inventor Descartes. His paper was read to the Royal Society of Edinburgh.

James also wrote poetry, but his mathematical talents were greater. He started at the University of Edinburgh at 16 and later continued his studies at the University of Cambridge, Britain's leading institution for mathematics. William Hopkins, who coached him for his exams, said that James was "the most extraordinary man I have ever met."

James earned his degree and remained at Cambridge as a postgraduate student, doing experiments on light. Then he read Faraday's *Experimental Researches* and started studying electricity. To cut a long story very short, he took Faraday's mechanical models of electromagnetic phenomena and by 1864 had distilled them into a system of four mathematical laws. (In the notation of the day there were more than four, but we now use vector notation to group them into four. Some formalisms reduce these down to one.) The laws describe electricity and magnetism in terms of two "fields," one electric and one magnetic, which pervade the whole of space. These fields describe not just the strength of electricity or magnetism at each location but the direction as well.

The four equations have simple physical meanings. Two tell us that electricity and magnetism can be neither created nor destroyed. The third describes how a time-varying magnetic field affects the surrounding electric field, and it embodies in mathematical form Faraday's discovery of induction. The fourth describes how a time-varying electric field affects the surrounding magnetic field. Even in words, these equations are elegant.

A simple mathematical manipulation of Maxwell's four equations confirmed something that Maxwell had long suspected: light is an electromagnetic wave, a propagating disturbance in the electric and magnetic fields.

The mathematical reason was that from Maxwell's equations it is easy to derive something that all mathematicians could recognize: the "wave equation," which as its name suggests describes how waves propagate. Maxwell's equations also predict the speed of such waves: they must travel at the speed of light.

Only one thing travels at the speed of light.

In those days it was assumed that waves had to be waves *in* something. There had to be a medium to transmit them; waves were vibrations of that medium. The obvious medium for light waves was the aether. The mathematics said that light waves had to vibrate at right angles to the direction of travel. This explained why Newton and Huygens had been so confused: they thought the waves vibrated along the direction of travel.

The theory made another prediction: that the "wavelength" of electromagnetic radiation, the distance from one wave to the next, could be *anything*. The wavelength of light is extremely short, but there ought to exist electromagnetic waves of much greater length. It was a good enough theory to inspire Heinrich Hertz to generate such waves, which we now call radio waves. Guglielmo Marconi quickly followed up with a practical transmitter and receiver, and suddenly we could talk to each other, almost instantly, across the entire planet. Now we send pictures the same way, monitor the skies with radar, and navigate with the Global Positioning System.

Unfortunately, the concept of the aether was problematic. If the aether existed, then the Earth, which revolves round the Sun, must be moving with respect to the aether. It ought to be possible to detect that motion— or else the very concept of the aether would have to be abandoned as inconsistent with experiment.

The answer to this conundrum would completely change the face of physics.

In the summer of 1876, the firm of Israel and Levi, run by two Jewish merchants in the city of Ulm in the state of Württemberg, gained a new partner, Hermann Einstein. In his youth, Hermann had shown considerable ability in mathematics, but his parents could not afford to send him to university. Now he was becoming a partner in a firm that sold featherbeds.

In August, Hermann married Pauline Koch in Cannstadt synagogue, and the couple eventually made a home in Bahnhofstrasse—Station Road. Less than eight months later, their first child was born. According to the birth certificate, "A child of the male sex, who has received the name Albert, was born in Ulm, in [Hermann's] residence, to his wife Pauline Einstein, née Koch, of the Israelitic faith." Five years later, Albert was presented with a sister, Maria, and the two became very close.

Albert's parents had a relaxed attitude to their religion and made efforts to integrate themselves into the regional culture. At that time, many German Jews were "assimilationist," toning down their cultural traditions so that they would fit in better with fellow citizens of other faiths. The names that Hermann and Pauline chose for their children were not traditional Jewish names, although they maintained that Albert was named "after" his grandfather Abraham. Religion was not a frequent topic of discussion in Hermann's house, and the Einsteins did not observe traditional Jewish rituals.

Maria's childhood recollections, published in 1924, are our main source of information about Albert's early experiences and personality. Apparently, he frightened his mother at birth because the back of his head was strangely angular and unusually large. "Much too heavy! Much too heavy!" she cried, when she first saw her baby. Fears that the boy would turn out to be mentally handicapped grew when it took him a long time to start to speak. But Albert was merely waiting until he was confident that he knew what he was doing. He later said that he only began to talk when he could master complete sentences. He would try them out in his head, and then utter them once he was sure the words were correct.

Albert's mother was an accomplished piano player. Between the ages of six and thirteen, Albert was given violin lessons from a teacher named Schmied. In later life, he was devoted to his violin, but in childhood he found the lessons boring.

The featherbed business having flopped, Hermann turned his hand to gas and water supplies, in collaboration with his brother Jakob. Jakob was an engineer and an entrepreneur, and the Einsteins invested heavily in the new venture. Then Jakob decided to diversify into electricity—not installing utilities but manufacturing equipment for power stations. The company officially came into being in 1885, and the two brothers moved into the same house in Munich, with financial help from Pauline's father and other family members. At first, the business did well, and the Elektronische Fabrik J. Einstein und Co. sold power stations in the Munich area and as far afield as Italy.

Einstein tells us that his interest in physics was triggered when his father showed him a compass. Then aged four or five, Albert was fascinated by its ability to point in the same direction no matter how it was turned, and he gained his first glimpse of the hidden wonders of the physical universe. He found the experience almost mystical.

At school, Albert was competent but initially showed no special brilliance. He was slow and methodical, received good grades, but was a poor mixer. He much preferred to play on his own; he was particularly fond of building houses of cards. He disliked sports. When he moved to the gymnasium in 1888 he developed a talent for Latin, and until he left at fifteen he always was at the top of his class in Latin and mathematics. His mathematical abilities were stimulated by Uncle Jakob, who as an engineer would have studied quite a bit of higher mathematics. Jakob would set young Albert mathematical problems, and Albert was delighted when he solved them. A family friend, Max Talmud, also had a significant effect on Albert's education. Talmud was a poverty-stricken medical student, and Hermann and Pauline had him over for dinner every Thursday evening. He gave Albert several books on popular science; then he initiated the young man into the philosophical writings of Immanuel Kant. The two would discuss philosophy and mathematics for hours. Talmud wrote that he never saw Einstein playing with other children, and that his reading material was always serious, nothing lightweight. His sole relaxation was to play music, including Beethoven and Mozart sonatas accompanied by Pauline.

Albert's enthusiasm for mathematics received a boost in 1891 when he acquired a copy of Euclid that he later called his "holy geometry book." What impressed him most was the clarity of the logic, the way Euclid had organized the flow of ideas. For a time, Albert became very devout, thanks to compulsory school instruction (in Catholicism, as it happened—there was no choice) and home tuition in the Jewish faith. But all this was brushed aside when he found out about science. His studies of Hebrew and his progress towards his bar mitzvah ground abruptly to a halt; Albert had found a different calling.

✳

By the early 1890s, all was not well in the Elektronische Fabrik J. Einstein und Co. Sales were becoming more difficult in Germany, and the company's Italian agent Lorenzo Garrone suggested that it should move to Italy. In June 1894, the German company was wound up, the family home went on the market, and the Einsteins moved to Milan—with the sole exception of Albert, who had his schooling to complete. While "Einstein and Garrone" set up shop in Pavia, where the family subsequently moved, Albert was left on his own in Munich.

It was a depressing experience, and he hated it. Not only that: the prospect of military service was looming. Without telling his parents, he decided to join them in Italy. He persuaded the family doctor to provide a certificate stating that he suffered from nervous disorders, which may well have been true; permitted to leave school early, he turned up unannounced in Pavia in the spring of 1895. His parents were horrified, so he promised to continue his studies so that he could take the entrance examination to ETH (the Eidgenossische Technische Hochschule, then as now a leading Swiss institution of higher education) in Zurich.

Albert blossomed in the Italian sunshine. In October he took the ETH entrance examination and failed. He passed easily in mathematics and science but fell down on the humanities. His essay writing was none too good either. But it turned out that there was another way into ETH, which was to start by gaining a high-school diploma, the Matura, which was an automatic entry route. He therefore went to a school in Aarau as a paying guest of the Winteler family. The Wintelers had seven children, and Albert enjoyed their company, developing a lasting affection for his substitute parents. He praised the school's "liberal spirit" and excellent teachers—saying pointedly that the teachers did not bow to outside authority.

For the first time in his life, he was happy at school. He grew in confidence and made his opinions known. One of his school essays, in French, laid out his plans for the future, which were to study mathematics and physics.

In 1896 he entered ETH, renouncing his Württemberg citizenship and becoming stateless. He saved one-fifth of his monthly allowance to pay for his eventual Swiss naturalization. But now the electrical factory owned by his father and uncle Jakob went bankrupt, taking much of the family fortune with it. Jakob took a regular job with a big company, but Hermann was determined to start yet another business. He ignored Albert's advice to the contrary, started again in Milan, and lasted only two years before that enterprise, too, failed. Albert once more became depressed by his family's misfortunes, until his father followed Jakob's lead and took a job installing power stations.

Albert spent much of his time at ETH in the physics laboratory, performing experiments. His professor, Heinrich Friedrich Weber, was unimpressed. "You are a smart boy, Einstein, a very smart boy," he told the young man. "But you have one great fault: you do not let yourself be told anything." He stopped Albert from carrying out an experiment to find out whether the Earth was moving relative to the aether—the hypothetical all-pervading fluid that was supposed to transmit electromagnetic waves.

Nor was Einstein greatly impressed by Weber, whose courses he found old-fashioned. He was especially disappointed not to be told more about Maxwell's theory of electromagnetism and taught it to himself, using a German text of 1894. He took lecture courses from two famous mathematicians, Hurwitz and Hermann Minkowski. Minkowski, a brilliantly original thinker, had introduced fundamental new methods into the theory of numbers, and was later to make important mathematical contributions to relativity. Albert also read some of Charles Darwin's works on evolution.

In order to proceed at ETH, he now needed to land an assistantship—what we would now call a teaching assistant position—so that he could finance his further studies while remaining at ETH. Weber hinted that he might offer Albert such a post, but failed to follow through, and Albert never entirely forgave him. He wrote a letter to Hurwitz inquiring whether such a post might be available, and apparently received a positive reply, but again nothing happened. By the end of 1900 he was unemployed. He did, however, publish his first research paper, on the forces acting between molecules. Soon thereafter, he attained Swiss citizenship, which he kept for the rest of his life, even after moving to the United States.

Throughout 1901, Albert kept trying to obtain a university position, writing letters, sending out copies of his paper, applying for any position that was open. No luck. In desperation he took a job as a temporary high-school teacher. To his surprise, he discovered that he enjoyed teaching; in addition, it left him ample spare time to continue his research into physics. He told his friend Marcel Grossmann that he was working on the theory of gases, and—once again—the motion of matter through the aether. He moved to another temporary teaching post in another school.

Now Grossmann came to Albert's rescue: Marcel's father was persuaded to recommend Albert to the director of the Federal Patent Office in Bern. When the job was officially advertised, Einstein applied. He resigned from school teaching and moved to Bern early in 1902, although he had not yet been told officially that he had secured the post. Perhaps he had been assured of this informally, or perhaps he was just very confident. The appointment was made official in June 1902. It was not the academic position that he coveted, but it earned enough money—3500 Swiss francs a year—to provide food, clothes, and lodging. And it left enough time for physics.

At ETH he had encountered a young student named Mileva Maric, who had a strong interest in science—and in Albert. They fell in love. Unfortunately, Pauline Einstein disliked her prospective daughter-in-law, and this caused ill feeling. Then Hermann developed terminal heart disease. On his deathbed the father finally agreed to allow Albert and Mileva to marry, but then he asked everyone in the family to leave him, so that he could die alone. Albert felt guilty for the rest of his life. He and Mileva were married in January 1903; their only son, Hans Albert, was born in May 1904.

The patent office job suited Einstein, and he carried out his duties so effectively that toward the end of 1904 his job was made permanent—but his boss warned that further promotion would depend on Einstein coming to grips with machine technology. His physics advanced too, with work on statistical mechanics.

All of which led up to the "golden year" of 1905, when the patent office clerk wrote a paper that eventually earned him the Nobel Prize. In the same year he obtained his PhD from the University of Zurich. He was also promoted to technical expert second class, with a raise of 1000 Swiss francs per year—it seems he had managed to master machine technology.

Even after he became famous, Albert always gave credit to Grossmann for paving the way to the job at the patent office. It was this, more than anything else, said Einstein, that had made his work in physics possible. It had been a stroke of genius, the perfect job, and he never forgot that.

In that most remarkable year in the history of physics, Einstein published three major research papers.

One was on Brownian motion, the random movements of very tiny particles suspended in a fluid. This phenomenon is named after its discoverer, the botanist Robert Brown. In 1827, he was looking through his microscope at grains of pollen floating in water. Inside holes in the pollen he noticed even tinier particles jiggling about at random. The mathematics of this kind of motion was worked out by Thorvald Thiele in 1880, and independently by Louis Bachelier in 1900. Bachelier's inspiration was not Brownian motion as such, but the equally random fluctuations of the stock market—the mathematics proved identical.

The physical explanation was still up for grabs. Einstein, and independently the Polish scientist Marian Smoluchowski, realized that Brownian motion might be evidence for the then-unproved theory that matter was made of atoms, which combined to form molecules. According to the so-called "kinetic theory," molecules in gases and liquids are constantly bouncing off each other, effectively moving at random. Einstein worked out enough of the mathematics of such a process to show that it matched the experimental observations of Brownian motion.

The second paper was on the photoelectric effect. Alexandre Becquerel, Willoughby Smith, Heinrich Hertz, and several others had observed that certain types of metal produce an electric current when exposed to light. Einstein started from the quantum-mechanical proposal that light is composed of tiny particles. His calculations showed that this assumption gives a very good fit to the experimental data. It was one of the first strong pieces of evidence in favor of quantum theory.

Either of these articles would have been a major breakthrough. But the third outclassed them all. It was on special relativity, the theory that went beyond Newton to revolutionize our views of space, time, and matter.

Our everyday view of space is the same as Euclid's and Newton's. Space has three dimensions, three independent directions at right angles to each other like the corner of a building—*north, east,* and *up.* The structure of space is the same at all points, though the matter that occupies space may vary. Objects in space can be moved in different ways: they can be rotated, reflected as if in a mirror, or "translated"—slid sideways without rotating. More abstractly, we can think of these transformations being applied to space itself (a change of the "frame of reference"). The structure of space, and the physical laws that express that structure and operate within it, are symmetric under these transformations. That is, the laws of physics are the same in all locations and at all times.

In a Newtonian view of physics, time forms another "dimension" that is independent of those of space. Time has a single dimension, and its symmetry transformations are simpler. It can be translated (add a fixed period of time to every observation) or reflected (run time in reverse—as a thought-experiment only). The physical laws do not depend on the starting date for your measurements, so they should be symmetric under translations of time. Most fundamental physical laws are also symmetric under time reversal, though not all, a fact that is rather mysterious.

But when mathematicians and physicists started to think about the newly discovered laws of electricity and magnetism, the Newtonian view seemed not to fit. The transformations of space and time that left the laws unchanged were not the simple "motions" of translation, rotation, and reflection; moreover, those transformations could not be applied to space or time independently. If you made a change in space alone, the equations got messed up. You had to change time in a compensating way.

To some extent this problem could be ignored, as long as the system under study was not moving. But the problem came to a head with the mathematics of a moving electric particle such as an electron—and this problem was central to the physics of the late nineteenth century. The associated worries about symmetry could no longer be ignored.

In the years leading up to 1905, a number of physicists and mathematicians had been puzzling about this strange feature of Maxwell's equations. If you performed an experiment involving electricity and magnetism in a laboratory or on a moving train, how should the results compare?

Of course, few experimentalists work on moving trains, but they all work on a moving *Earth*. For many purposes, though, the Earth can be considered to be at rest, because the experimental apparatus moves along with it, so the motion makes no real difference. Newton's laws of motion, for example, remain exactly the same in any "inertial" frame of reference, one that is moving with constant speed in a straight line. The Earth's speed is fairly constant, but it spins on its axis and revolves around the Sun, so the motion relative to the Sun is not straight. Still, the path the apparatus follows is almost straight; whether the curvature matters depends on the experiment, and often it does not matter at all.

No one would have been worried if Maxwell's equations had to take a different form in a rotating frame. What they discovered was more disturbing: Maxwell's equations took a different form in an inertial frame. Electromagnetism on a moving train is different from electromagnetism in a fixed laboratory, even when the train is traveling in a straight line at constant speed.

There was a further complication: it is all very well to say that a train, or the Earth, is moving, but the concept of motion is relative. Mostly we don't notice the movement of the Earth, for example. The Sun's rising in the morning and setting in the evening is *explained* by the Earth's rotation. But we don't *feel* the rotation, we deduce it.

If you sit in a train and look out of the window, you may get the impression that you are fixed and the countryside is rushing past you. Someone standing in a field watching you go past observes the opposite: she is stationary and the train is moving. When we say that the Earth goes around the Sun rather than the Sun going around the Earth, we are making a subtle distinction, because either description is valid, depending on which frame of reference you choose. If the frame is carried along with the Sun, then the Earth moves relative to that frame and the Sun does not. But if the frame is carried along with the Earth, as the planet's inhabitants are, then the Sun is the object that moves.

So what was all the fuss about the heliocentric theory, which holds that the Earth orbits the Sun, not the other way around? Poor Giordano Bruno was burnt to death because he said that one description was correct while the Church preferred the other one. Did he die because of a misunderstanding?

Not exactly. Bruno made a number of claims that the Church viewed as heresies—small matters like the nonexistence of God. His fate would have been much the same if he had never mentioned the heliocentric theory. But there is an important sense in which "the Earth goes around the Sun" is superior to "the Sun goes around the Earth." The important difference is that the mathematical description of the planets' movements relative to the Sun is much simpler than that of their movements relative to the Earth. An Earth-centered theory is possible but very complicated. Beauty is more significant than mere truth. Many points of view yield true descriptions of nature, but some provide more insight than others.

Now, if all motion is relative, then nothing can be absolutely "at rest." Newtonian mechanics is consistent with the next-simplest proposal: that all inertial frames are on the same footing. But that is not true of Maxwell's equations.

※

As the nineteenth century drew to a close, one further intriguing possibility also had to be considered. Since light was believed to be a wave traveling through the aether, then perhaps the aether was at rest. Instead of all motions being relative, some motions—those relative to the aether—might be *absolute*. But that still did not explain why Maxwell's equations are not the same in all inertial frames.

The common theme here is symmetry. Changing from one frame of reference to another is a symmetry operation on space-time. Inertial frames are about translational symmetries; rotating frames are about rotational symmetries. Saying that Newton's laws are the same in any inertial frame is to say that those laws are symmetric under translations at uniform speed. For some reason, Maxwell's equations do not have this property. That seems to suggest that some inertial frames are more inertial than others. And if any inertial frames are special, surely it should be those that are stationary relative to the aether.

The upshot of these problems, then, was two questions, one physical, one mathematical. The physical one was, can motion relative to the aether be detected in experiments? The mathematical one was, what are the symmetries of Maxwell's equations?

The answer to the first was found by Albert Michelson, a US Navy officer who was taking leave to study physics under Helmholtz, and the chemist Edward Morley. They built a sensitive device to measure tiny discrepancies in the speed of light moving in different directions, and concluded that there were no discrepancies. Either the Earth was at rest relative to the aether—which made little sense given that it was circling the Sun—or there was no aether, and light did not obey the usual rules for relative motion.

Einstein attacked the problem from the mathematical direction. He didn't mention the Michelson–Morley experiment in his papers, though he later said he was aware of it and that it had influenced his thinking. Instead of appealing to experiments, he worked out some of the symmetries of Maxwell's equations, which have a novel feature: they mix up space and time. (Einstein did not make the role of symmetry explicit, but it is not far below the surface.) One implication of these weird symmetries is that uniform motion relative to the aether—assuming that such a medium exists—cannot be observed.

Einstein's theory acquired the name "relativity," because it made unexpected predictions about relative motion and electromagnetism.

※

"Relativity" is a very bad name. It's misleading because the most significant feature of Einstein's theory is that some things are *not* relative. Specifically, the speed of light is absolute. If you shine a beam of light past an observer standing in a field, and another one standing in a moving train, both will measure the *same* speed.

This is distinctly counterintuitive, and at first sight it seems absurd. The speed of light is roughly 186,000 miles per second. Clearly this is what the observer in the field should measure. What about the person on the train?

Suppose the train is traveling at 50 mph. First, imagine that there is a second train on a parallel line, also traveling at 50 mph. You look out of the window and watch it go past. How fast does it seem to you to be moving?

If it is traveling in the same direction that you are, then the answer is 0 mph. The second train will keep pace with yours, it will stay alongside it, and seem not to be moving relative to your train. If it is traveling the opposite way, then it will appear to flash past at 100 mph, because your train's 50 mph is in effect added to the speed of the oncoming train.

If you do the measurements with trains, that is what you find.

Now replace the second train by a beam of light. The speed of light, converted to the appropriate units, is 670,616,629 miles per hour. If your train were moving away from the source of the light, you would expect to observe a speed of 670,616,629 - 50 = 670,616,579 mph, because the light would have to "catch up" with the train. On the other hand, if your train were moving toward the source of the light, then you would expect the speed of light relative to the train to be 670,616,629 + 50 = 670,616,679 mph, because the movement of the train would add to the apparent speed.

According to Einstein, both of those numbers are wrong. What you will observe, in either case, is light traveling at 670,616,629 mph—exactly the same speed that the woman in the field observes.

This sounds mad. If the Newtonian rules for relative motion work for another train, why don't they work for light? Einstein's answer is that laws of physics are different from Newton's for objects that move very fast.

More precisely, the laws of physics are different from Newton's, period. But the difference only becomes apparent when objects are moving at speeds very close to the speed of light. At low speeds like 50 mph, Newton's laws are such a good *approximation* to Einstein's proposed replacements that you cannot notice any difference. But as speeds increase, discrepancies become large enough to be observed.

The basic physical point is that the symmetries of the Maxwell equations not only preserve the *equations*; they preserve the speed of light. Indeed, the speed of light is built into the equations. So the speed of light must be absolute.

It is ironic that this proposal should be called "relativity." Einstein actually wanted to name it "Invariantentheorie": *invariant* theory. But the name "relativity" stuck, and in any case there already existed an area of mathematics called invariant theory, so Einstein's preferred name might have been confusing. Though not half as confusing as using "relativity" to describe the invariance of the speed of light in all inertial frames.

The consequences of "relativity" are bizarre. The speed of light is a limiting speed. You can't travel faster than light, and you can't send messages faster than light. No *Star Wars* hyperdrives. Near the speed of light, lengths shrink, time slows to a crawl, and mass increases without limit. But—and here's the wonderful thing—you don't notice, because your measuring instruments *also* shrink, slow down (in the sense that time passes more slowly), or get heavier. This is why the observer in the field and the one on the train measure your light at the same speed despite their relative motion: the changes in length and time compensate exactly for the expected effects of the relative movement. This is why Michelson and Morley could not detect the Earth's motion relative to the aether.

When you are moving, everything looks the same to you as it did when you weren't moving. The laws of physics cannot tell you whether you are moving or stationary. They can tell you whether you are accelerating, but not how fast you are going if your speed is constant.

It may still seem weird, but experiments confirm the theory in exquisite detail. Another consequence is Einstein's famous formula $E = mc^2$, linking mass to energy, which indirectly led to the atomic bomb, though its role there is often exaggerated.

Light is so familiar to us that we seldom think about how weird it is. It seems to weigh nothing, it penetrates everywhere, and its enables us to see. What is light? Electromagnetic waves. Waves in what? The space-time continuum, which is a fancy way of saying, "we don't know." Early in the twentieth century, the medium for the waves was thought to be the luminiferous aether. After Einstein, we understood one thing about that aether: it doesn't exist. The waves are not *in* anything.

Quantum mechanics, as we will see, went further. Not only are light waves not *in* anything: all things *are* waves. In place of a medium to support the waves—a fabric of space-time that ripples as the waves pass—the fabric *itself* is made of waves.

Einstein was not the only person to notice that the symmetries of space-time, as revealed in Maxwell's equations, are not the obvious Newtonian symmetries. In a Newtonian view, space and time are separate and different. Symmetries of the laws of physics are combinations of rigid motions of space and an independent shift in time. But as I mentioned, these transformations do not leave Maxwell's equations invariant.

Pondering this, the mathematicians Henri Poincaré and Hermann Minkowski were led to a new view of the symmetries of space and time, on a purely mathematical level. If they had described these symmetries in physical terms, they would have beaten Einstein to relativity, but they avoided physical speculations. They did understand that symmetries of the laws of electromagnetism do not affect space and time independently but mix them up. The mathematical scheme describing these intertwined changes is known as the Lorentz group, after the physicist Hendrik Lorentz.

Minkowski and Poincaré viewed the Lorentz group as an abstract expression of certain features of the laws of physics, and descriptions like "time passing more slowly" or "objects shrinking as they speed up" were thought of as vague analogies rather than anything real. But Einstein insisted that these transformations have a genuine physical meaning. Objects, and time, really do behave like that. He was led to formulate a physical theory, special relativity, that incorporated the mathematical scheme of the Lorentz group into a physical description not of space and a separate time, but of a unified space-time.

Minkowski came up with a geometric picture for this non-Newtonian physics, now called Minkowski space-time. It represents space and time as independent coordinates, and a moving particle traces out a curve—which Einstein called its *world line*—as time passes. Because no particle can travel faster than light, the slope of the world line can never get more than 45° away from the time direction. The particle's past and future always lie inside a double cone, its light cone.

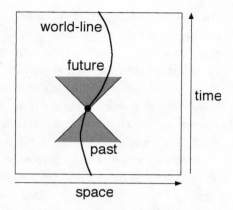

Geometry of Minkowski space-time.

That took care of electricity and magnetism, two basic forces of nature. But one basic force was still missing from this description: *gravity*. Attempting to develop a more general theory that included gravity, and again relying on the principle that the laws of nature must be symmetric, Einstein was led to general relativity: the idea that space-time itself is curved and that its curvature corresponds to mass. From these ideas emerged our current cosmology of the Big Bang, in which the universe grew from a tiny speck some 13 billion years ago, and the remarkable concept of a black hole, an object so massive that light cannot escape its gravitational field.

General relativity traces back to early work on non-Euclidean geometry, which led Gauss to the idea of a "metric," a formula for the distance between any two points. New geometries arise when this formula is not the classic Euclidean one derived from the Pythagorean theorem. As long as the formula obeys a few simple rules, it defines a meaningful concept of "distance." The main rule is that the distance from one point A to another C cannot decrease if you pass through a point B in between. That is, the direct distance from A to C is less than or equal to the distance from A to B plus that from B to C. This is the "triangle inequality," so called because in Euclidean geometry it states that any side of a triangle is shorter than the other two put together.

The Pythagorean formula holds in Euclidean geometry, in which space is "flat." So when the metric is different from the Euclidean one, we can attribute that difference to some kind of "curvature" of space. You can visualize this as a bending of space, but that's not really the best picture because there must then be a bigger space for the original one to bend *in*. A better way to think of "curvature" is that regions of space are either compressed or stretched, so that from the inside they seem to hold less, or more, space than they do from outside. (Fans of the British TV series *Doctor Who* will be reminded of the Tardis, a spaceship/time-machine whose inside is larger than its outside.) Gauss's brilliant student Riemann extended the idea of a metric from two dimensions to any number, and he modified the idea so that distances can be defined locally—just for points that are very close together. Such a geometry is called a Riemannian manifold, and it is the most general kind of "curved space."

Physics happens not in space but in space-time, where—according to Einstein—the natural "flat" geometry is not Euclidean but Minkowskian. Time enters into the "distance" formula in a different way from space. Such a geometric setup is a "curved space-time." It turned out to be just what the patent clerk ordered.

✳

Einstein struggled a long time to devise his equations for general relativity. He first investigated how light moves in a gravitational field, and this led him to base his later research on a single fundamental principle, the *equivalence principle*. In Newtonian mechanics, gravity has the effect of a force, pulling bodies toward each other. Forces cause accelerations. The equivalence principle states that accelerations are always indistinguishable from the effects of a suitable gravitational field. In other words, the way to put gravity into relativity is to understand accelerations.

By 1912, Einstein had convinced himself that a theory of gravity cannot be symmetric under every Lorentz transformation; that kind of symmetry applies *exactly, everywhere,* only when matter is absent, gravity is zero, and space-time is Minkowskian. By abandoning this requirement of "Lorentz-invariance" he saved himself a lot of fruitless effort. "The only thing I believed firmly," he wrote in 1950, "was that one had to incorporate the equivalence principle in the fundamental equations." But he also recognized the limitations even of that principle: it should be valid only locally, as a kind of infinitesimal approximation to the true theory.

By 1907, Einstein's friend Grossmann had become a geometry professor at ETH, and Albert was persuaded to take a position there too. Not for long—after a year he left for Berlin and later went to Prague. But he kept in contact with Grossmann, and this paid off handsomely. In 1912, Grossmann helped Einstein to work out what kind of mathematics he should be thinking about:

> This problem remained unsolvable to me until . . . I suddenly real-ized that Gauss's theory of surfaces held the key for unlocking the mystery . . . However, I did not know at that time that Riemann had studied the foundations of geometry in an even more profound way . . . My dear friend the mathematician Grossman was there when I returned from Prague to Zurich. From him I learned for the first time about Ricci and later about Riemann. So I asked my friend whether my problem could be solved by Riemann's theory.

"Ricci" is Gregorio Ricci-Curbastro, the coinventor, along with his stu-dent Tullio Levi-Civita, of calculus on Riemannian manifolds. The Ricci tensor is a measure of curvature, simpler than Riemann's original concept.

Other sources have Einstein saying to Grossmann, "You must help me, or else I'll go crazy!" Grossmann delivered. As Einstein later wrote, he "not only saved me the study of the relevant mathematical literature, but also supported me in the search for the field equations of gravita-tion." In 1913, Einstein and Grossmann published the first fruits of their combined labors, ending with a conjecture about the form of the required field equations: the stress-energy tensor must be proportional to . . . something.

What?

They didn't yet know. It had to be another tensor, another measure of curvature.

At that point they both made mathematical errors, which set them off on a lengthy wild goose chase. They were convinced, correctly, that their theory had to yield Newtonian gravity in a suitable limiting case—flat space-time, small gravity. They deduced from this some technical con-straints on the sought-for equation, that is, constraints on the nature of the required "something." But their arguments were fallacious and the constraints did not apply.

Einstein was convinced that the correct field equations should deter-mine the mathematical form of the metric—the distance formula in space-time, which determines all of its geometrical properties—uniquely.

This is simply wrong: changes in the coordinate system can change the formula while having no effect on the *intrinsic* curvature of the space. But Einstein was unaware of the so-called Bianchi identities, which clarify the lack of uniqueness, and apparently so was Grossmann.

It was every researcher's nightmare: an apparently watertight idea, which seemed to lead in the right direction but was actually leading them up the garden path. Eradicating such mistakes is desperately hard, because you're convinced they are not mistakes. Often you don't even realize what assumptions you are tacitly making.

At the end of 1914, Einstein finally realized that the field equations cannot determine the metric uniquely because of the possibility of choosing a different coordinate system, which has no physical implications but changes the formula for the metric. He still did not know the Bianchi identities, but now he didn't need them. He finally knew that he was free to choose whichever coordinates were most convenient.

On 18 November 1914, Einstein opened up a new front in his war with the gravitational field equations. He had gotten close enough to his final formulation to start making predictions. He made two. One—really a "postdiction," made after the event—explained a tiny change already observed in the orbit of the planet Mercury. The "perihelion" position, where the planet comes closest to the Sun, was slowly changing. Einstein's new theory of gravity told him how fast the perihelion should be moving—and his calculation was spot on.

The second prediction required new observations to verify or falsify it—which was excellent news, because new observations are the best tests of new theories. According to Einstein's theory, gravity should bend light. The geometry of this effect is simple, and it concerns geodesics—the shortest path between any two points. If you stretch a string tight and hold it in mid-air, it forms a straight line; this happens because in Euclidean space a straight line is a geodesic. If, however, you hold the two ends of the string against a football and pull it tight, it forms a curve lying on the surface of the ball. Geodesics on a curved space—the ball—are themselves curved. The same happens in a curved space-time, though the details are slightly different.

✳

The physical circumstances in which this effect might show up are also straightforward. A star, such as the Sun, will bend any light that passes nearby. The only way to observe this effect, at that time, was to wait for an eclipse of the Sun, when the Sun's light no longer drowned out the light from stars whose position in the sky was close to the Sun's edge. If Einstein was right, the apparent positions of those stars should shift slightly, compared to their positions when they were not aligned with the Sun.

The quantitative analysis of this phenomenon is less straightforward. Einstein's first attempt, in 1911, predicted a shift of just under a second of arc. Newton would have predicted a similar amount, based on his belief that light is made from tiny particles: the force of gravity would attract the particles, causing their paths to bend. But by 1915 Einstein had deduced that in his new theory, the light should bend by twice that amount, 1.74 seconds of arc.

Now there was a real prospect of deciding between Newton and Einstein. On 25 November 1914, Einstein wrote down his field equations in their final form. These *Einstein equations* constitute the basis of general relativity, the relativistic theory of gravity. They are written in a mathematical formalism known as a *tensor*—a kind of hyped-up matrix. Einstein's equation tells us that the Einstein tensor is proportional to the rate of change of the stress-energy tensor. That is, the curvature of space-time is proportional to the quantity of matter present. These equations obey a kind of symmetry principle, but it is a local one. In small regions of space-time, they have the same symmetries as special relativity, provided the local effect of curvature is taken into account.

Einstein noted that his calculations of the motion of the perihelion of Mercury and the deflection of light by a star remained unchanged by the minor modifications that he had made. He presented his equations to the Prussian Academy, only to discover that the mathematician David Hilbert had already submitted the identical equations but had claimed far more for them than just a theory of gravity. In fact, he had claimed that they included the electromagnetic equations, which was a mistake. It is fascinating to see, yet again, a top mathematician coming very close to beating Einstein to the punch.

Several attempts were made to verify Einstein's prediction that light would be deflected by the gravitational field of the Sun. The first, in Brazil, was spoiled by rain. In 1914, a German expedition went to observe an eclipse in Crimea, but when World War I began they were instructed to return home—fast. Some did. The others were arrested but eventually made their way home unharmed. Naturally, no observations were made.

The war prevented observations in Venezuela in 1916. The Americans tried in 1918, with inconclusive results. Finally, a British expedition led by Arthur Eddington succeeded in May 1919, but did not announce its results until November.

When it did, the verdict favored Einstein over Newton. There *was* a deflection, it was too big to fit a Newtonian model, and it fitted Einstein's beautifully.

In retrospect, the experiments were not as decisive as they seemed. The range of experimental error was quite large, and the best conclusion was that Einstein was *probably* right. (More recent observations, with better techniques and equipment, have confirmed Einstein's theory.) But at the time, they were represented as definitive, and the media went ape. Anyone who could prove Newton wrong must be a genius. Anyone who could discover radically new physics must be the greatest living scientist.

Thus was a legend born. Einstein wrote about his ideas in the *Times* of London. A few days later, the paper's editorial page responded:

This is news distinctly shocking and apprehensions for the safety of confidence even in the multiplication table will arise . . . It would take the presidents of two Royal Societies to give plausibility or even thinkability to the declaration that light has weight and space has limits. It just doesn't by definition, and that's the end of that—for commonfolk, however it may be for higher mathematicians.

But the higher mathematicians were right. Soon the *Times* was telling the world that "only twelve people can understand the theory of 'the suddenly famous Dr. Einstein,'" a myth that circulated for years, even when large numbers of physics undergraduates were routinely being taught the theory in their coursework.

In 1920, Grossmann showed the first symptoms of multiple sclerosis. He wrote his last paper in 1930 and died in 1936. Einstein went on to become the iconic physicist of the twentieth century. In later life he grew to tolerate his fame, finding it vaguely amusing. Early on, he seems to have enjoyed interacting with the media.

But now we must leave Einstein's career, except to remark that after 1920 his efforts in physics were devoted to a fruitless quest to combine relativity and quantum mechanics in a single "unified field theory." He was still working on this problem the day before his death, in 1955.

12

A QUANTUM QUINTET

"Almost everything is already discovered, and all that remains is to fill a few holes." This is discouraging news for a talented young man intending to study physics, especially when the news comes from someone who ought to know: in this case, Philipp von Jolly, a physics professor.

The date was 1874, and von Jolly's view reflected what most physicists of the period believed: physics was done. In 1900 no less a luminary than Lord Kelvin said, "There is nothing new to be discovered in physics now. All that remains is more and more precise measurement."

Mind you, he also said, "I can state flatly that heavier than air flying machines are impossible," and "Landing on the moon offers so many serious problems for human beings that it may take science another 200 years to lick them." Kelvin's biographer wrote that he spent the first half of his career being right and the second half being wrong.

But he wasn't totally wrong. In his 1900 lecture "Nineteenth-Century Clouds over the Dynamical Theory of Heat and Light," he put his finger on two crucial gaps in the period's understanding of the physical universe: "The beauty and clearness of the dynamical theory, which asserts heat and light to be modes of motion, is at present obscured by two clouds. The first involves the question, How could the Earth move through an elastic solid, such as essentially is the luminiferous ether? The second is the Maxwell–Boltzmann doctrine regarding the partition of energy." The first cloud led to relativity, the second to quantum theory.

Fortunately, the young recipient of Jolly's advice was not daunted. He had no wish to discover *new* things, he said—all he wanted was to develop a better understanding of the known fundamentals of physics. In the

search for this understanding, he brought about one of the two great revolutions in twentieth-century physics, and dispelled Kelvin's second cloud.
His name was Max Planck.

Julius Wilhelm Planck was a professor of law in Kiel and Munich. His father and his mother had both been theology professors, and his brother
was a judge. So when his second wife, Emma Patzig, presented Julius with
a son—his sixth child—the boy was certain to grow up in an intellectual
environment. Max Karl Ernst Ludwig Planck was born on 23 April 1858.
Europe was in the usual political turmoil, and the boy's earliest memories
included Prussian and Austrian troops marching into Kiel during the
Danish–Prussian War of 1864.

By 1867 the Plancks had moved to Munich, and Max was being tutored
by the mathematician Hermann Müller at the King Maximilian School.
Müller taught the boy astronomy, mechanics, mathematics, and some basic physics, including the law of conservation of energy. Planck was an excellent student, and he graduated unusually early, at the age of sixteen.

He was also a talented musician, but he decided to study physics despite
Jolly's well-intentioned advice. Planck carried out some experiments under
Jolly's supervision but quickly switched to theoretical physics. He kept
company with some of the world's leading physicists and mathematicians,
moving to Berlin in 1877 to study under Helmholtz, Gustav Kirchhoff,
and Weierstrass. He passed his first examinations in 1878 and obtained a
doctorate in 1879 with a thesis on thermodynamics. For a time he taught
mathematics and physics at his old school. In 1880, his habilitation thesis,
on equilibrium states of bodies at different temperatures, was accepted,
and he was qualified for a permanent academic career. He duly secured
such a position, but not until 1885, when the University of Kiel made him
an associate professor. His research focused on thermodynamics, especially the concept of entropy.

Max met Marie Merck, the sister of a friend, and in 1887 they married
and rented an apartment. In all, they had four children: Karl, twins Emma
and Grete, and Erwin.

In 1889, the year the twins were born, Max was appointed to Kirchhoff's position in Berlin, becoming a full professor in 1892. The family
moved to a villa in the Grunewald region of Berlin, close to a number of
other leading academics. One, the theologian Adolf von Harnack, became

a close friend. The Plancks were sociable, and famous intellectuals visited their house regularly. These included Einstein and the physicists Otto Hahn and Lise Meitner, who later made fundamental discoveries about nuclear fission, part of the long development leading to the atomic bomb and nuclear power stations. At these events the Plancks continued a tradition of playing music, started by Helmholtz.

For a time life was rosy, but Marie contracted a lung disease, possibly tuberculosis, and died in 1909. A year and a half later, at 52, Max remarried, this time to Marga von Hoesslin, with whom he had a third son, Hermann.

In 1894, a local electrical company was trying to develop a more efficient light bulb, so Max started some industrial contract research. Theoretically, the analysis of a light bulb was a standard physics problem known as "blackbody radiation"—how light would be emitted by a perfectly nonreflective body. Such a body, when heated, emits light of all frequencies, but the intensity of the light, or equivalently its energy, varies with the frequency. A fundamental question was, how does the frequency affect the intensity? Without such basic data, it would be difficult to invent a better light bulb.

There were good experimental results, and one theoretical law, the Rayleigh–Jeans law, had been derived from basic principles of classical physics. Unfortunately, this law disagreed with experiment at high frequencies. In fact, it predicted something impossible: as the light's frequency increases, its energy should become infinitely large. This impossible result became known as the "ultraviolet catastrophe." Further experiments led to a new law, which fitted the observations for high-frequency radiation, known as Wien's law after its discoverer, Wilhelm Wien.

However, Wien's law went wrong for *low*-frequency radiation.

Physicists were faced with two laws: one working at low frequencies but not at high ones, the other doing the exact opposite. Planck hit on the idea of interpolating between the two: that is, writing down a mathematical expression that approximated the Rayleigh–Jeans law at low frequencies and Wien's law at high frequencies. The resulting formula is now called the Planck law for blackbody radiation.

This new law was deliberately designed to match experiments beautifully, across the entire spectrum of electromagnetic radiation, but it was purely empirical—derived from experiments, not from any basic physical principle. Planck, pursuing his avowed intention to understand known physics better, was dissatisfied, and he devoted much effort searching for physical principles that would lead to his formula.

Eventually, in 1900, Planck noticed a curious feature of his formula. He could derive it by much the same calculation that Rayleigh and Jeans had employed, provided he made one tiny change. The classical derivation had assumed that for any given frequency, the energy of electromagnetic radiation could in principle take any value whatsoever. In particular, it could get as close to zero as you wished. Planck realized that this assumption was the cause of the ultraviolet catastrophe, and that if he made a different assumption, that troublesome infinity disappeared from the calculation.

The assumption, though, was radical. The energy of radiation of a given frequency had to come as a whole number of "packets" of fixed size. In fact, the size of each packet had to be proportional to the frequency—that is, equal to the frequency multiplied by some constant, which we now call Planck's constant and write using the symbol h.

These energy packets were called *quanta* (singular: *quantum*). Planck had quantized light.

All very well, but why had experimentalists never noticed that the energy was always a whole number of quanta? By comparing his calculations with the observed energies, Planck was able to calculate the size of his constant, and it turned out to be very, very small. In fact, h is roughly 6×10^{-34} joule-seconds. Roughly speaking, to notice the "gaps" in the possible range of energies—the values that classical physics permitted but quantum physics did not—you had to make observations that were accurate to the 34th decimal place. Even today, very few physical quantities can be measured to more than six or seven decimal places, and in those days three was asking a lot. Direct observation of quanta required absurd levels of accuracy.

It may seem strange that a mathematical difference so tiny that it can never be seen could have such a huge effect on the radiation law. But the calculation of the law involves adding up all the contributions to the energy from all possible frequencies. The result is a collective effect of all possible quanta. From the Moon you can't spot an individual grain of sand on Earth. But you can see the Sahara. If sufficiently many very tiny units combine, the result can be huge.

Planck's physics thrived, but his personal life was filled with tragedy. His son Karl was killed in action during the First World War. His daughter Grete died in childbirth in 1917, and Emma suffered the same fate in 1919, having married Grete's widower. Much later, Erwin was executed by the Nazis for taking part in the unsuccessful 1944 attempt to assassinate Adolf Hitler.

✳

By 1905, new evidence had turned up that supported Planck's radical proposal, in the form of Einstein's work on the photoelectric effect. Recall that this is the discovery that light can be converted into electricity. Einstein was aware that electricity comes in discrete packages. Indeed, by then physicists knew that electricity is the motion of tiny particles called electrons. From the photoelectric effect, Einstein deduced that the same must be true of light. This not only verified Planck's ideas about light quanta, it explained what the quanta are: light waves, like electrons, must be particles.

How can a wave be a particle? Yet that was the unequivocal message of the experiments. The discovery of particles of light, or photons, quickly led to the quantum picture of the world in which particles are really waves, behaving sometimes like one, sometimes like the other.

The physics community started to take quanta more seriously. The great Danish physicist Niels Bohr came up with a quantized model of the atom, in which electrons move in circular orbits around a central nucleus, with the size of the circle being limited to discrete quanta. The French physicist Louis de Broglie reasoned that since photons can be both waves and particles, and electrons are emitted by suitable metals when they are impacted by photons, then electrons must *also* be both waves and particles. Indeed, all matter must possess this dual existence—sometimes solid particle, sometimes undulating wave. That's why experiments can indicate either form.

Neither "particle" nor "wave" really describes matter at extremely tiny scales. The ultimate constituents of matter are a bit of both—*wavicles*. De Broglie invented a formula to describe wavicles.

Now came a key step, essential to our story. Erwin Schrödinger took de Broglie's formula and turned it into an equation that describes how wavicles move. Just as Newton's laws of motion were fundamental to classical mechanics, Schrödinger's equation became fundamental to quantum mechanics.

Erwin was born in Vienna in 1886, the offspring of a mixed marriage. His father, Rudolf Schrödinger, manufactured cerecloth, a waxy cloth used to make shrouds for the dead; he was also a botanist. Rudolf was a Catholic, while Erwin's mother, Georgine Emilia Brenda, was a Lutheran. From 1906 to 1910 Erwin studied physics in Vienna under Franz Exner and Friedrich Hasenöhrl, becoming Exner's assistant in 1911. He gained his habilitation in 1914, at the start of World War I, and spent the war as an officer in the Austrian artillery. Two years after the war ended, he married Annemarie Bertel. In 1920 he became the equivalent of an associate professor in Stuttgart, and by 1921 he was a full professor in Breslau, now the city Wrocław, in Poland.

He published the equation that is now named after him in 1926, in a paper showing that it gives the correct energy levels for the spectrum of the hydrogen atom. This was quickly followed by three other major papers on quantum theory. In 1927, he joined Planck in Berlin, but in 1933, upset by the anti-Semitism of the Nazis, he left Germany for Oxford, where he was made a fellow of Magdalen College. Not long after he arrived, he and Paul Dirac were awarded the Nobel Prize in physics.

Schrödinger maintained a scandalously unorthodox lifestyle, living with two women, and this offended the tender sensibilities of the Oxford dons. Within a year he had moved again, this time to Princeton, where he was offered a permanent position, but he decided not to accept—possibly because his attachment to both wife and mistress in the same household didn't go down any better in Princeton than it did in Oxford. Eventually, he settled in Graz, Austria, in 1936, and ignored the opinions of strait-laced Austrians.

Hitler's occupation of Austria caused severe difficulties for Schrödinger, a known Nazi opponent. He publicly rejected his earlier views (and much later apologized to Einstein for doing so). The ploy didn't work: he lost his job because he was politically unreliable, and had to flee to Italy.

Schrödinger finally settled in Dublin. The year 1944 saw the publication of *What Is Life?* an intriguing but flawed attempt to apply quantum physics to the problem of living creatures. He based his ideas on the concept of "negentropy," the tendency of life to disobey—or somehow subvert—the second law of thermodynamics. Schrödinger emphasized that the genes of living creatures must be some kind of complicated molecule, containing coded instructions. We now call this molecule DNA, but its structure was discovered only in 1953, by Francis Crick and James Watson—inspired, in part, by Schrödinger.

In Ireland, Schrödinger retained his relaxed attitude to sexuality, getting involved with students and fathering two children by different mothers. He died of tuberculosis in Vienna in 1961.

Schrödinger is best known for his cat. Not a real cat, but one that appeared in a thought experiment. It is generally interpreted as a reason for not considering Schrödinger's waves to be real physical things. Instead, they are thought of as a behind-the-scenes description that can never be verified experimentally but that has the right consequences. However, this interpretation is controversial—if the waves do not exist, why do their consequences all work out so nicely?

Anyway, back to the cat. According to quantum mechanics, wavicles can interfere with each other, piling up on top of each other and reinforcing when peak meets peak, and canceling each other out when peak meets trough. This type of behavior is called "superposition," so quantum wavicles can superpose—implying that they can contain a variety of potential states without fully existing in any of them. Indeed, according to Bohr and the famous "Copenhagen interpretation" of quantum theory, that is the natural state of affairs. Only when we observe some physical quantity do we force it out of some quantum superposition and into a single "pure" state.

This interpretation works well for electrons, but Schrödinger wondered what it would imply for a cat. In his thought experiment, a cat locked in a box can be in a superposition of the states alive and dead. When you open the box, you observe the cat and force it into either one state or the other. As Pratchett noticed in *Maskerade,* cats aren't like that. Greebo, a hyper-macho cat, emerges from a box in a third state: absolutely bloody furious.

Schrödinger also knew that cats aren't like that, though for different reasons. An electron is a submicroscopic entity, and it behaves like something on the quantum level. It possesses (when we measure it) a particular position or velocity or spin that can be described relatively simply. A cat is macroscopic, and it doesn't. You can superpose electron states, but not cats. My wife and I have two cats, and when they try to superpose, the result is flying fur and two highly indignant cats. The jargon term here is "decoherence," which explains why large quantum systems like cats look like the familiar "classical" systems in our daily lives. Decoherence tells us that the cat contains so many wavicles that they all get tangled up together and ruin the superposition in less time than light can travel the diameter of an electron. So cats, being macroscopic systems composed of an absolutely gigantic number of quantum particles, *behave* like cats. They can be alive, or dead, but not both at once.

Nonetheless, on suitably small scales—and we are talking very small stuff here, not anything you can see in a normal microscope—the universe behaves just as quantum physics says it does, and it can do two different things at the same time. And that changes everything.

Just how strange the quantum world must be emerged from the research of Werner Heisenberg. Heisenberg was a brilliant theoretical physicist, but his grasp of experiments was so poor that during his examination for the doctorate he couldn't answer simple questions about telescopes and microscopes. He didn't even know how a battery worked.

August Heisenberg married Anna Wecklein in 1899. He was a Lutheran, she a Catholic, and she converted to his religion to make the marriage possible. They had much in common: he was a teacher and an expert classicist specializing in ancient Greek, while she was a head teacher's daughter and an expert on the Greek tragedies. Their first son, Erwin, was born in 1900 and became a chemist. Their second, Werner, was born in 1901 and changed the world.

Germany was still a monarchy at this time, and the teaching profession carried high social status, so the Heisenbergs were financially comfortable and could send their sons to good schools. In 1910, August was made professor of medieval and modern Greek at the University of Munich, to which city the family moved. In 1911, Werner started at the King Maximilian School in Munich, where Planck had also studied. Werner's grandfather, Nikolaus Wecklein, was the school principal. The boy was bright and quick, partly because his father encouraged him to compete with his elder brother, and showed remarkable abilities in math and science. He had musical talent too, and learned the piano so well that by the age of 12 he was performing in school concerts.

Heisenberg later wrote that "both my interests in languages and in mathematics were awakened rather early." He earned top grades in Greek and Latin and did well in mathematics, physics, and religion. His worst subjects were athletics and German. His mathematics teacher, Christoph Wolff, was excellent, and stretched Werner's abilities by setting him special problems to solve. Soon the pupil had outclassed the teacher, and Heisenberg's school report stated, "With his independent work in the mathematical-physical field he has come far beyond the demands of the school." He taught himself relativity, preferring its mathematical content to its physical implications. When his parents asked him to tutor a local college student for her exams, he taught himself calculus, a subject not included in the school curriculum. He developed an interest in number theory, saying that "it's clear, everything is so that you can understand it to the bottom."

To help Werner improve his Latin, his father brought him some old papers on mathematics, written in that language. Among them was Kronecker's dissertation on a topic ("complex units") in algebraic number theory. Kronecker, a world-class number theorist, famously believed that "God created the integers—all else is the work of Man." Heisenberg was inspired to have a go at proving Fermat's Last Theorem. After nine years in the school he graduated at the top of his class and attended the University of Munich.

When World War I broke out, the Allies blockaded Germany. Food and fuel were in very short supply; the school had to be closed because it could not be heated, and on one occasion Werner was so weak from starvation that he fell off his bike into a ditch. His father and his teachers were fighting in the army; the young men who remained behind received military training and nationalistic indoctrination. The end of the war brought the end of the German monarchy as well, and Bavaria briefly had a socialist government along Soviet lines, but in 1919 German troops from Berlin kicked out the socialists and restored a more moderate social democracy.

Like most of his generation, Werner was disillusioned by Germany's defeat and blamed his elders for their military failure. He became the leader of a group associated with the New Boy Scouts, an extremist right-wing organization that aimed to restore the monarchy and dreamed of a Third Reich. Many branches of the New Boy Scouts were anti-Semitic, but Werner's group included a number of Jewish boys. He spent a lot of time with his boys, camping and hiking and generally trying to recapture a romantic vision of Germany as it once had been, but these activities ended in 1933 when Hitler banned all youth organizations other than those he had set up himself.

In 1920, Werner went to the University of Munich, intending to become a pure mathematician until an interview with one of the pure math professors put him off the idea. He decided instead to study physics under Arnold Sommerfeld. Immediately recognizing Werner's abilities, Sommerfeld allowed him to attend advanced classes. Soon Werner had done some original research on the quantum approach to atomic structure. His doctorate was awarded in 1923, breaking the university's record for speed. In the same year, Hitler tried to overthrow the Bavarian government in the "beer hall putsch," intended as a prelude to a march on Berlin, but the attempt failed. Hyperinflation was rampant; Germany was coming to pieces.

Werner continued working. He collaborated with many leading physicists, all of whom were thinking about quantum theory because that was where the action was. He worked with Max Born to devise a better theory of the atom. It occurred to Heisenberg to represent the state of an atom in terms of the frequencies observed in its spectrum—the kinds of light that it emitted. He boiled this idea down to a peculiar kind of mathematics involving lists of numbers. Born eventually realized that this kind of list was actually quite respectable: mathematicians called it a matrix. Happy that the ideas made sense, Born sent the paper off for publication. As the ideas developed, they matured into a new, systematic mathematics of quantum theory: matrix mechanics. It was seen as a competitor to Schrödinger's wave mechanics.

⁂

Who was right? It turned out the two theories were identical, as Schrödinger discovered in 1926. They were two distinct mathematical representations of the same underlying concepts—just as Euclidean methods and algebra are two equivalent ways of looking at geometry. At first Heisenberg could not believe this, because the essence of his matrix approach was the existence of discontinuous jumps as an electron changed its state. The entries in his matrices were the associated changes of energy. He couldn't see how waves, as continuous entities, could model discontinuities. In a letter to the Austrian-Swiss physicist Wolfgang Pauli, he wrote, "The more I think about the physical portion of Schrödinger's theory, the more repulsive I find it . . . What Schrödinger writes about the visualizability of his theory 'is probably not quite right,' in other words, it's crap." But really this disagreement was a rerun of a much older debate, in which Bernoulli and Euler had disagreed about solutions of the wave equation. Bernoulli had a formula for the solutions, but Euler could not see how this formula, which looked continuous, could represent discontinuous solutions. Nevertheless, Bernoulli was right, and so was Schrödinger. His *equations* might be continuous, but many features of their solutions could be discrete—including the energy levels.

Most physicists preferred the wave-mechanical picture because it was more intuitive. Matrices were a bit too abstract. Heisenberg still preferred his lists, because they consisted of observable quantities, and it seemed impossible to detect one of Schrödinger's waves experimentally. In fact, the Copenhagen interpretation of quantum theory, dramatized as Schrödinger's cat, stated that any attempt to do so would "collapse" the wave into a single, well-defined spike. So Heisenberg became more and more concerned about what aspects of the quantum world can be measured, and how. You can measure every entry in his lists. You can't do that for one of Schrödinger's waves. Heisenberg considered this difference a powerful reason for sticking to matrices.

Following this line of thought, he discovered that in principle you can measure a particle's position as accurately as you wish—but there is a price to be paid, because the more accurately you know the position, the less accurately you can know the momentum. Conversely, if you can measure the momentum very accurately, you lose track of the position. The same trade-off occurs for energy and time. You can measure one or the other, but not both. Not if you want high-accuracy measurements.

This wasn't a problem with experimental procedure; it was an inherent feature of quantum theory. He wrote out his reasoning in a letter to Pauli in February 1927. The letter eventually inspired a paper, and Heisenberg's idea acquired the name "uncertainty principle." It was one of the first examples of an inherent limitation in physics. Einstein's assertion that nothing can move faster than light was another.

In 1927, Heisenberg became Germany's youngest professor, at the University of Leipzig. In 1933, the year Hitler rose to power, Heisenberg won the Nobel Prize for physics. This made him a highly influential figure, and his willingness to stay in Germany during the Nazi regime made many believe that Heisenberg was himself a Nazi. As far as can be established, he wasn't. But he was a patriot, and that led him to associate with Nazis and be complicit in many of their activities. There is some evidence that Heisenberg tried to stop the ruling powers kicking Jews out of university positions, but to no effect. In 1937, he found himself described as a "white Jew" and was under threat of being sent to a concentration camp, but after a year he was cleared of suspicion by Heinrich Himmler, head of the SS. Also in 1937, Heisenberg married Elisabeth Schumacher, the daughter of an economist. Their first children were twins; eventually they had seven.

During World War II, Heisenberg was one of the leading physicists involved in Germany's search for nuclear weapons—the "atomic bomb." He worked on nuclear reactors in Berlin, while his wife and children were dispatched to the family's summer home in Bavaria. His role in Germany's atom bomb project has proved highly controversial. When the war ended he was detained by the British and held for six months for questioning in a country house near Cambridge. The transcripts of his interrogations, recently made public, have exacerbated the controversy. Heisenberg does say at one point that he was solely interested in making a nuclear reactor ("engine") and did not want to be involved in a bomb. "I would say that I was absolutely convinced of the possibility of our making a uranium engine, but I never thought we would make a bomb, and at the bottom of my heart I was really glad that it was to be an engine and not a bomb. I must admit that." The truth of this claim is still hotly debated.

After the war, and his release from British custody, Heisenberg went back to work on quantum theory. He died of cancer in 1976.

✳

Most of the great German creators of quantum theory came from an intellectual background—they were the sons of doctors, lawyers, academics, or other professionals. They lived in expensive homes, played music, and took part in the local social life and culture. The great English creator of quantum mechanics had a very different and much sadder childhood, with an autocratic and distinctly eccentric father who was largely estranged from his own parents and family, and a mother who was so browbeaten that she and two of her children ate in the kitchen while her husband and their younger son ate in total silence in the dining room.

The father was Charles Adrien Ladislas Dirac, born in the Swiss canton of Valais in 1866, who ran away from home at the age of 20. Charles arrived in Bristol in 1890 but did not become a British citizen until 1919. In 1899, he married Florence Hannah Holten, a sea captain's daughter, and their first child, Reginald, was born the next year. Two years later a second son, Paul Adrien Maurice, was added to the growing family; four years after that they had a daughter, Beatrice.

Charles did not tell his parents that he had married, or that they had become grandparents, until 1905, when he visited his mother in Switzerland. By then, his father had been dead for ten years.

Charles worked as a teacher at the Merchant Venturer's Technical College in Bristol. He was generally considered a good teacher, but he was also renowned for a total absence of human feelings and very strict discipline. He was, in short, a martinet, but so were many teachers.

Paul, a natural introvert, was made even more so by his father's curious isolation and lack of any social life. Charles insisted that Paul speak to him only in French, presumably to encourage him to learn that language. Since Paul's French was dreadful, he found it simpler not to speak at all. Instead, he spent his time wondering about the natural world. The antisocial dining arrangements in the Dirac household also seem to have stemmed from the rule that conversation should be held entirely in French. It was never clear whether Paul actively hated his father or just disliked him, but when Charles died, Paul's main comment was "I feel much freer now."

Charles was proud of Paul's intellectual abilities and very ambitious for his children—by which he meant that they should do what he had planned for them. When Reginald said he wanted to become a doctor, Charles insisted that he become an engineer. In 1919, Reginald obtained a very poor engineering degree; five years later, while working on an engineering project in Wolverhampton, he killed himself.

Paul lived at home with his parents, and also studied engineering at the same college as his brother. His favorite subject was mathematics, but he chose not to study that. Possibly he didn't want to go against his father's wishes; but he was also under the erroneous impression, still widespread today, that the only career for someone with a mathematics degree is school-teaching. No one had told him that there were alternatives—among them, research.

Salvation came in the form of a newspaper headline. The front page of the *Times* for 7 November 1919 shrieked, REVOLUTION IN SCIENCE.

NEW THEORY OF THE UNIVERSE. NEWTONIAN IDEAS OVER-
THROWN. Halfway down the second column was the subheading SPACE
"WARPED." Suddenly everyone was talking about relativity.

Recall that one of the predictions of general relativity is that gravity
bends light, by twice the amount that Newton's laws would predict. Frank
Dyson and Sir Arthur Stanley Eddington had mounted an expedition to
Príncipe Island in West Africa, where a total eclipse of the Sun was due.
Simultaneously, Andrew Crommelin, of Greenwich Observatory, led a
second expedition to Sobral, in Brazil. Both parties observed stars near
the edge of the Sun during the period of totality, and found slight dis-
placements in the stars' apparent positions, consistent with Einstein's predictions, but not with Newtonian mechanics.

Einstein, an overnight celebrity, sent his mother a postcard: "Dear
Mother, joyous news today. H. A. Lorentz telegraphed that the English
expeditions have actually demonstrated the deflection of light from the
Sun." Dirac was hooked. "I was caught up in the excitement produced by
relativity. We discussed it very much. The students discussed it among
themselves, but had very little accurate information to go on." Public
knowledge of relativity was confined largely to the word; philosophers
claimed that they had known for years that "everything is relative," and
dismissed the new physics as old hat. Unfortunately, they only displayed
their ignorance, and the ease with which they had fallen for misleading
terminology.

Paul went to some lectures on relativity by Charlie Broad, then a philosophy professor at Bristol, but their mathematical content was insignificant.
Eventually, he bought a copy of Eddington's *Space, Time and Gravitation*
and taught himself the necessary mathematics and physics. Before leaving
Bristol, he knew both special and general relativity inside out.

Paul was good at theory, terrible at laboratory work. In later years, physicists spoke of the "Dirac effect": he had only to walk into a laboratory for
nearby experiments to go wildly wrong. An engineering profession would
have been a disaster. He found himself with a first-class degree, but unemployed at a time when jobs were scarce because of the postwar economic
depression. Luckily, he was offered the chance to study mathematics at
Bristol University, all tuition paid, and he leaped at it. There he specialized
in applied mathematics.

In 1923, Paul became a postgraduate research student at the University of Cambridge, where his shyness was a real handicap. He wasn't interested in sports, made few friends, and had nothing whatsoever to do with women. He spent most of his time in the library. In 1920, he had spent the summer working at the same factory as his brother Reginald. The two would often pass in the street but never stopped to talk, so ingrained was the habit of silence between family members.

Paul quickly rose to prominence; within six months he had written his first research paper. Other papers followed in a rapid stream. Then, in 1925, he encountered quantum mechanics. On a long autumn walk in the Cambridgeshire countryside he found himself thinking about Heisenberg's "lists." These are matrices, and matrices do not commute, something that had initially bothered Heisenberg. Dirac was aware of Lie's idea that in these circumstances the important quantity is the commutator $AB - BA$, not the product AB, and he was struck by the intriguing thought that a very similar concept occurs in the Hamiltonian formalism of mechanics, where it is called a Poisson bracket. But Dirac couldn't remember the formula.

The thought kept him awake much of the night, and the next morning he "hurried along to one of the libraries as soon as it was open, and then I looked up Poisson brackets in Whittaker's *Analytical Dynamics,* and I found that they were just what I needed." His discovery was this: the commutator of two quantum matrices is equal to the Poisson bracket of the corresponding classical variables, multiplied by the constant $ih/2\pi$. Here h is Planck's constant, i is $\sqrt{-1}$, and π is, well, π.

This was a dramatic discovery. It told physicists how to turn classical mechanical systems into quantum ones. The mathematics was very beautiful, linking two deep but previously unconnected theories. Heisenberg was impressed.

Dirac's contributions to quantum theory are many, and I will select just one of the high points, his relativistic theory of the electron, which dates from 1927. By then, the quantum theorists knew that electrons have "spin"—somewhat analogous to the spin of a ball about an axis, but with strange features that make the analogy very rough. If you take a spinning ball and rotate the system through a full 360°, both ball and spin get back to where they started. But when you do the same to an electron, the spin *reverses.* You have to rotate through 720° before the spin gets back to its original value.

This is actually very similar to quaternions, whose interpretation as "rotations" of space has the same quirk. Mathematically, rotations of space form the group SO(3), but the relevant group for both quaternions and electrons is SU(2). These groups are almost the same, but SU(2) is twice as big, built—in a certain sense—out of two copies of SO(3). It is called a "double cover," and the result is to expand a 360° rotation into one through twice that angle.

Dirac didn't use quaternions, and he didn't use groups either. But over the Christmas season at the end of 1927 he came up with "spin matrices," which play the same role. Mathematicians later generalized Dirac's matrices to "spinors," which are important in the representation theory of Lie groups.

The spin matrices allowed Dirac to formulate a relativistic quantum model of the electron. It did everything he had hoped for—and a little more. It predicted solutions with negative energy as well as the expected ones with positive energy. Eventually, after some false starts, this puzzling feature led Dirac to the concept of "antimatter"—that every particle has a corresponding antiparticle, with the same mass but the opposite charge. The antiparticle of the electron is the positron, and it was unknown until Dirac predicted it.

The laws of physics remain (almost) unchanged if you replace every particle with its antiparticle—so that operation is a symmetry of the natural world. Dirac, who was never terribly impressed by group theory, had discovered one of the most fascinating symmetry groups in nature.

From 1935 onward, until his death in Tallahassee in 1984, Dirac placed enormous value on the mathematical elegance of physical theories, and used that principle as a touchstone for his research. If it wasn't beautiful, he believed, it was wrong. When he visited Moscow State University in 1956, and followed tradition by writing words of wisdom on a blackboard to be kept for posterity, he wrote, "A physical law must possess mathematical beauty." And he talked of a "mathematical quality" in nature. Yet he never seemed to think of group theory as beautiful, possibly because physicists mostly approach groups through massive calculations. Only mathematicians, it seemed, were attuned to the exquisite beauty of Lie groups.

Beautiful or not, group theory soon became essential reading for any budding quantum theorist, thanks to the son of a leather merchant.

At the turn of the nineteenth century, leather was big business, and it still is. But in those days a small businessman could make a good living by tanning and selling leather. A good example was Antal Wigner, the director of a tannery. He and his wife Erzsébet were of Jewish descent, but did not practice Judaism. They lived in what was then Austro-Hungary, in the city of Pest. Conjoined with neighboring Buda, this town became today's Budapest, the capital of Hungary.

Jenő Pál Wigner, the second of their three sons, was born in 1902, and between the ages of five and ten he was educated at home by a private tutor. Soon after Jenő started school, he was diagnosed with tuberculosis and sent to an Austrian sanatorium to recover. He stayed there for six weeks before it turned out that the diagnosis had been wrong. Had it been right, he would almost certainly not have survived to adulthood.

Made to lie on his back much of the day, the boy did mathematical problems in his head to pass the time. "I had to lie on a deck chair for days on end," he later wrote, "and I worked terribly hard on constructing a triangle if the three altitudes are given." The altitudes of a triangle are the three lines that pass through a corner and hit the opposite side at right angles. Finding the altitudes, given the triangle, is easy. Going in the opposite direction is decidedly more difficult.

After Jenő left the sanatorium he continued to think about mathematics. In 1915, at Budapest's Lutheran High School, he met another boy who would become one of the world's leading mathematicians: Janós (later John) von Neumann. But the two never became more than loose acquaintances, because von Neumann tended to keep to himself.

In 1919, the Communists overran Hungary and the Wigners fled to Austria, returning to Budapest later in the same year when the Communists were kicked out. The entire family converted to Lutheranism, but this had little effect on Jenő, he later said, because he was "only mildly religious."

In 1920, Jenő finished school near the top of his class. He wanted to become a physicist, but his father wanted him to join the family leather-tanning business. So instead of taking a physics degree, Jenő studied chemical engineering, which his father thought would help to advance the business. For his first year at university, the young man went to the Budapest Technical Institute; then he switched to the Technische Hochschule in Berlin. He ended up spending most of his time in the chemical laboratory, which he enjoyed, and precious little time in theoretical classes.

Still, Jenő had not given up on physics. The University of Berlin was not far away, and who might be there but Planck and Einstein, along with lesser luminaries. Jenő took advantage of their proximity and went to the immortals' lectures. He completed his doctorate, with a thesis on the formation and breakup of molecules, and duly joined the tannery. Predictably, this proved a bad idea: "I did not get along very well in the tannery . . . I did not feel at home there . . . I did not feel that this was my life." His life was mathematics and physics.

In 1926, he was contacted by a crystallographer at the Kaiser Wilhelm Institute who wanted a research assistant. The duties would combine both of Wigner's interests, in a chemical context. The project had a huge influence on Wigner's career, and thus on the course of nuclear physics, because it introduced him to group theory—the mathematics of symmetry. The first major application of group theory to physics had been the classification of all 230 possible crystal structures. Wigner wrote, "I received a letter from a crystallographer who wanted to find out why the atoms occupy positions in the crystal lattices which correspond to symmetry axes. He also told me that this had to do with group theory and that I should read a book on group theory and then work it out and tell him."

Perhaps no less dismayed than his son by Jenő's foray into the tanning trade, Antal Wigner agreed to allow the research assistantship. Jenő started by reading a few of Heisenberg's papers on quantum theory, and developed a theoretical method to calculate the spectrum of an atom with three electrons. But he also realized that his methods would prove extraordinarily complicated for more electrons than three. At this point, he turned for advice to his old acquaintance von Neumann, who suggested he read about group representation theory. This area of mathematics was heavily laden with the algebraic concepts and techniques of the time, notably matrix algebra. But thanks to his studies in crystallography and his familiarity with a leading algebra textbook of the period—Heinrich Weber's *Lehrbuch der Algebra*—matrices posed no problem for Wigner.

Von Neumann's advice proved sound. If an atom possesses some number of electrons, then since all electrons are identical, the atom does not "know" which electron is which. In other words, the equations describing the radiation emitted by that atom must be symmetric under all permutations of those electrons. Using group theory, Wigner developed a theory of the spectrum of atoms with any number of electrons.

To that point, his work had taken place within the traditional realm of classical physics. But quantum theory was where the excitement was. Now he embarked on his life's masterwork, the application of group representation theory to quantum mechanics.

Ironically, he did so despite, not because of, his new job. David Hilbert, the elder statesman of German mathematics, had developed a keen interest in the mathematical principles behind quantum theory and required the services of a research assistant. In 1927, Wigner went to Göttingen to join Hilbert's research group. His ostensible role was to provide physical insight to inform Hilbert's vast mathematical expertise.

It didn't quite work out as planned. The two met only five times in the course of a year. Hilbert was old, tired, and increasingly reclusive. So Wigner went back to Berlin, gave lectures on quantum mechanics, and continued to put together his most famous book: *Group Theory and Its Application to the Quantum Mechanics of Atomic Spectra.*

He had been partly anticipated by Hermann Weyl, who had also written a book about groups in quantum theory. But Weyl's main focus was on foundational issues, whereas Wigner wanted to solve specific physical problems. Weyl was after beauty, Wigner was seeking truth.

We can understand Wigner's approach to group theory in a simple, classical context, the vibrations of a drum. Musical drums are usually circular, but in principle they can be any shape. When you hit a drum with a stick, the skin vibrates and makes a noise. Different shapes of drum produce different sounds. The range of frequencies a drum can produce, called its *spectrum,* depends in a complex manner on the drum's shape. If the drum is symmetrical, we might expect the symmetry to show up in the spectrum. It does, but in a subtle way.

Imagine a rectangular drum—you don't see these often outside of mathematics departments. The typical patterns of vibration for such a drum divide it into a number of smaller rectangles, for example:

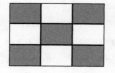
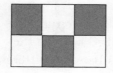

Two patterns of vibration of a rectangular drum.

Here we see two different vibrational patterns with two different frequencies. The pictures are snapshots of the patterns, taken at one instant. The dark regions are displaced downward, the white ones upward.

The symmetries of the drum have implications for the patterns, because any symmetry transformation of the drum can be applied to a possible pattern of vibration to produce another possible pattern of vibration. So the patterns come in symmetrically related sets. However, individual patterns need not have the same symmetries as the drum. For instance, a rectangle is symmetric under rotation through 180°. If we apply this symmetry transformation to the two patterns above, they become:

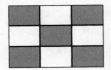 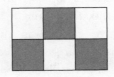

The same two patterns after rotating the drum through 180°.

The left-hand pattern is unchanged, so it shares the rotational symmetry of the drum. But the right-hand pattern has swapped dark regions for light. This effect is called *spontaneous symmetry-breaking,* and it is very common in physical systems: it occurs when a symmetric system has less-symmetric states. The left-hand pattern does not break symmetry, but the right-hand one does. Let's focus on the right-hand pattern and see what effect its broken symmetry has.

Although the pattern and its rotation are different, they both vibrate at the same frequency, because rotation is a symmetry of the drum and hence of the equations that describe its vibrations. So the spectrum of the drum contains this particular frequency "twice." It may seem difficult to detect that effect experimentally, but if you make small changes in the drum that destroy its rotational symmetry—say by making a small indentation along one edge—then the two frequencies drift slightly apart, and you can spot that there are two of them, very close together. This would not have happened if the frequency had occurred only once for the symmetric drum.

Wigner realized that the same effect arises with symmetric molecules, atoms, and atomic nuclei. The sounds made by the drum become vibrations of the molecules, and the spectrum of sounds is replaced by the spectrum of emitted or absorbed light. In the quantum world, the spectrum is created by transitions between different energy states, and the atom emits photons whose energy—hence frequency, thanks to Planck—

corresponds to that difference. Now the spectrum can be detected using a spectroscope. Again, some of the frequencies—observed as spectral lines—may be double (or multiple) because of the symmetry of the molecule, atom, or nucleus.

How can we detect this multiplicity? We can't make an indentation in the molecule, as we did for the drum. But we can place the molecule in a magnetic field. This also destroys the underlying symmetry and splits the spectral lines. Now you can use group theory—more strictly, group representation theory—to calculate the frequencies and how they split.

Representation theory is one of the most beautiful and powerful mathematical theories, but it is also technically demanding and full of hidden pitfalls. Wigner turned it into a high art. Others struggled to follow his lead.

By 1930, Wigner had secured a part-time post in America, at the Institute for Advanced Study, and he shuttled between Princeton and Berlin. In 1933, the Nazis passed laws forbidding Jews to hold university jobs, so Wigner moved permanently to the United States—mainly at Princeton, where he anglicized his name to Eugene Paul. His sister Margit joined him in Princeton. There she met Dirac, who was visiting, and in 1937 the two were married, to everyone's amazement.

Margit's marriage worked out fine, but Eugene's job did not. In 1936, Wigner wrote, "Princeton dismissed me. They never explained why. I could not help feeling angry." Actually, Wigner resigned, apparently because he was not advancing sufficiently quickly. Presumably he believed that Princeton's refusal to promote him had effectively forced him to resign, so he felt as though he had been fired.

He quickly found a new job at the University of Wisconsin, took US citizenship, and met a physics student named Amelia Frank. They married, but Amelia had cancer and died within a year.

At Wisconsin, Wigner turned his attention to nuclear forces and discovered that they are governed by the symmetry group SU(4). He also made a basic discovery concerning the Lorentz group, published in 1939. But group theory was not then a standard part of a physicist's training, and its main application was still to the rather specialized area of crystallography. To most physicists, group theory looked complicated and unfamiliar, a fatal combination. The quantum physicists, appalled by what was invading their patch, described the development as the "Gruppenpest," or "group disease." Wigner had triggered an epidemic and his colleagues did not want to catch it. But Wigner's views were prophetic. Group-theoretic methods came to dominate quantum mechanics, because the influence of symmetry is all-pervasive.

In 1941, Wigner began his second marriage, to a teacher named Mary Annette. They had two children, David and Martha. During the war, Wigner, like von Neumann and a great many top mathematical physicists, worked on the Manhattan Project to construct an atomic bomb. He was awarded the Nobel Prize in Physics in 1963.

Despite living for years in the USA, Wigner always longed for his homeland. "After 60 years in the United States," he wrote in his declining years, "I am still more Hungarian than American. Much of American culture escapes me." He died in 1995. The physicist Abraham Pais described him as "a very strange man . . . one of the giants of 20th century physics." The viewpoint he developed is revolutionizing the twenty-first century as well.

13

THE FIVE-DIMENSIONAL MAN

By the late twentieth century, physics had made extraordinary advances. The large-scale structure of the universe seemed to be very well described by general relativity. Remarkable predictions such as the existence of black holes—regions of space-time from which light can never escape, created by the collapse of massive stars under their own gravity—were supported by observations. The small-scale structure of the universe, on the other hand, was described in extraordinary detail and with exquisite precision by quantum theory, in its modern form of quantum field theory, which incorporates special but not general relativity.

There were two serpents in the physicist's paradise, however. One was a "philosophical" serpent: these two wildly successful theories disagreed with each other. Their assumptions about the physical world were mutually inconsistent. General relativity is "deterministic"—its equations leave no room for randomness. Quantum theory has inherent indeterminacy, captured by Heisenberg's uncertainty principle, and many events, such as the decay of a radioactive atom, happen at random. The other serpent was "physical": the quantum-based theories of elementary particles left a number of important issues unresolved—such as why particles have particular masses or indeed, why they have mass at all.

Many physicists believed that both serpents could be expelled from their Garden of Eden by the same bold action: *unify* relativity and quantum theory. That is, devise a new theory, a logically consistent one, that agrees with relativity on large scales and with quantum theory on small scales. This was what Einstein had tried to do for half his life—and failed. With typical modesty, physicists christened this unified view a "Theory of

Everything." The hope was that the whole of physics could be boiled down to a set of equations simple enough to be printed on a T-shirt.

It wasn't such a wild idea. You can certainly get Maxwell's equations on a T-shirt, and I currently own one with the equations of special relativity, with the slogan "Let there be light" in Hebrew. A friend bought it for me in the Tel Aviv airport. Less frivolously, major unifications of apparently disparate physical theories have been achieved before. Maxwell's theory united magnetism and electricity, once thought to be entirely distinct natural phenomena powered by entirely different forces of nature, into a single phenomenon: electromagnetism. The name may be awkward, but it accurately reflects the process of unification. A more modern instance, less well known except to the physics community, is the electroweak theory, which unified electromagnetism with the weak nuclear force—see below. A further unification with the strong nuclear force has left just one thing missing from the mix: gravity.

Given this history, it is entirely reasonable to hope that this final force of nature can be brought into line with the rest of physics. Unfortunately, gravity has awkward features that make this process difficult.

It could be that no Theory of Everything is possible. Although mathematical equations—"laws of nature"—have so far been very successful as explanations of our world, there is no guarantee that this process must continue. Perhaps the universe is less mathematical than physicists imagine.

Mathematical theories can approximate nature very well, but it is not certain that any piece of mathematics can capture reality *exactly*. If not, then a patchwork of mutually inconsistent theories might provide workable approximations valid in different domains—and there might not be a single overriding principle that combines all of those approximations and works in all domains.

Except, of course, for the trivial list of if/then rules: "If speeds are small and scales are big, use Newtonian mechanics; if speeds are large and scales are big use special relativity," and so on. Such a mix-and-match theory is horribly ugly; if beauty is truth, then mix-and-match can only be false. But perhaps at root the universe *is* ugly. Perhaps there is no root to be *at*. These are not appealing thoughts, but who are we to impose our parochial aesthetic on the cosmos?

The view that a Theory of Everything *must* exist brings to mind monotheist religion—in which, over the millennia, disparate collections of gods and goddesses with their own special domains have been replaced by *one* god whose domain is everything. This process is widely viewed as an advance, but it resembles a standard philosophical error known as "the equation of unknowns" in which the same cause is assigned to all mysterious phenomena. As the science fiction writer Isaac Asimov put it, if you are puzzled by flying saucers, telepathy, and ghosts, then the obvious explanation is that flying saucers are piloted by telepathic ghosts. "Explanations" like this give a false sense of progress—we used to have three mysteries to explain; now we have just one. But the one new mystery conflates three separate ones, which *might well have entirely different explanations.* By conflating them, we blind ourselves to this possibility.

When you explain the Sun by a sun-god and rain by a rain-god, you can endow each god with its own special features. But if you insist that both Sun and rain are controlled by the *same* god, then you may end up trying to force two different things into the same straitjacket. So in some ways fundamental physics is more like fundamentalist physics. Equations on a T-shirt replace an immanent deity, and the unfolding of the consequences of those equations replaces divine intervention in daily life.

Despite these reservations, my heart is with the physical fundamentalists. I would like to see a Theory of Everything, and I would be delighted if it were mathematical, beautiful, and true. I think religious people might also approve, because they could interpret it as proof of the exquisite taste and intelligence of their deity.

Today's quest for a Theory of Everything has its roots in an early attempt to unify electromagnetism and general relativity—at the time, the whole of known physics. This attempt was made only fourteen years after Einstein's first paper on special relativity, eight years after his prediction that gravity could bend light, and four years after the finished theory of general relativity was revealed to a waiting world. It was such a good attempt that it could easily have diverted physics onto a new course entirely, but unfortunately for its inventor, his work coincided with something that *did* set physics on a new course: quantum mechanics. In the ensuing gold rush, physicists lost interest in unified field theories; the world of the quantum offered far richer pickings, with far more chance of making a major discovery. It would be sixty years before the idea behind that first attempt was revived.

It began in the city of Königsberg, then the capital of the German province of East Prussia. Königsberg is now Kaliningrad, the administrative center of a Russian exclave lying between Poland and Lithuania. This city's surprising influence on the development of mathematics began with a puzzle. Königsberg lay on the river Pregel (now Pregolya), and seven bridges linked the two banks of the river to each other and to two islands. Did there exist a route that would permit the citizens of Königsberg to walk across every bridge in turn, never crossing the same bridge twice? One of those citizens, Leonhard Euler, developed a general theory of such questions, implying that in this case the answer was no, and thereby took one of the first steps toward the area of mathematics now called topology. Topology is about geometrical properties that remain unchanged when a shape is bent, twisted, squashed, and generally deformed in a continuous manner—no tearing or cutting.

Topology has become one of the most powerful developments in today's mathematics, with many applications to physics. It tells us the possible shapes of multidimensional spaces, a growing theme both in cosmology and particle physics. In cosmology we want to know the shape of space-time on the largest scale, that of the entire universe. In particle physics we want to know the shape of space and time on small scales. You might think the answer is obvious, but physicists no longer do. And their doubts also trace back to Königsberg.

In 1919, Theodor Kaluza, an obscure mathematician at the University of Königsberg, had a very strange idea. He wrote it up and sent it to Einstein, who apparently was struck speechless. Kaluza had found a way to combine gravity and electromagnetism in a single coherent "unified field theory," something that Einstein had been trying to do for many years without success. Kaluza's theory was very elegant and natural. There was just one disturbing feature: the unification required space-time to have *five* dimensions, not four. Time was the same as always, but space had somehow acquired a fourth dimension.

Kaluza had not set out to unify gravity and electromagnetism. For some reason best known to him, he had been messing around with five-dimensional gravity, a kind of mathematician's warmup exercise, working out how Einstein's field equations would look if space had that absurd extra dimension.

In four dimensions the Einstein equations have ten "components"—they boil down to ten separate equations describing ten separate numbers. These numbers jointly constitute the metric tensor, which describes the curvature of space-time. In five dimensions there are fifteen components, hence fifteen equations. Ten of them reproduce Einstein's standard four-dimensional theory, which is no surprise; four-dimensional space-time is embedded in five-dimensional space-time, so you would naturally expect the four-dimensional version of gravity to be embedded in the five-dimensional one. What about the remaining five equations? They could have been just some peculiar structure with no significance for our own world. But they weren't. Instead, they were very familiar, and that's what amazed Einstein. Four of Kaluza's remaining equations were precisely Maxwell's equations for the electromagnetic field, the ones that hold in *our* four-dimensional space-time.

The one remaining equation described a very simple kind of particle, which played an insignificant role. But no one, least of all Kaluza, had expected both Einstein's theory of gravity and Maxwell's theory of electromagnetism to emerge spontaneously from the five-dimensional analogue of gravity alone. Kaluza's calculation seemed to be saying that light is a vibration in an extra, hidden dimension of space. You could put gravity and electromagnetism together into a seamless whole, but only by supposing that space is really four-dimensional and space-time is five-dimensional.

Einstein agonized over Kaluza's paper, because there was absolutely no reason to imagine that space-time has an extra dimension. But eventually he decided that however strange the idea might seem, it was so beautiful, and potentially so far-reaching, that it should be published. After dithering for two years, Einstein accepted Kaluza's paper for a major physics journal. Its title was "On the unity of problems of physics."

All this talk of extra dimensions probably sounds rather vague and mystical. It is a concept associated with Victorian spiritualists, who invoked the fourth dimension as a convenient place to hide everything that didn't make sense in the familiar three. Where do spirits live? In the fourth dimension. Where does ectoplasm come from? The fourth dimension. Theologians even placed God and His angels there until they realized that the fifth was better and the sixth better still, and that finally only an infinite dimension would really do for an omniscient and omnipresent entity.

All great fun, but not terribly scientific. So it is worth digressing to clarify the underlying mathematics. The main point is that the "dimension" of some mathematical or physical setup is the number of distinct variables needed to describe it.

Scientists spend a lot of time thinking about variables—quantities that are subject to change. Experimental scientists spend even more time measuring them. "Dimension," which is just a geometric way to refer to a variable, has turned out to be so useful that it is now built into science and mathematics as a standard way of thinking and is considered to be entirely prosaic and unremarkable.

Time is a nonspatial variable, so it provides a *possible* fourth dimension, but the same goes for temperature, wind speed, or the lifespan of termites in Tanzania. The position of a point in three-dimensional space depends on three variables—its distances east, north, and upward relative to some reference point, using negative numbers for the opposite directions. By analogy, anything that depends on four variables lives in a four-dimensional "space," and anything that depends on 101 variables lives in a 101-dimensional space.

Any complex system is inherently multidimensional. The weather conditions in your backyard depend on temperature, humidity, three components of wind velocity, barometric pressure, intensity of rainfall—that's seven dimensions already, and there are plenty of others we might include. I bet you didn't realize you had a seven-dimensional backyard. The state of the nine (well, eight; alas, poor Pluto!) planets in the solar system is determined by six variables for each planet—three positional coordinates and three components of velocity. So our solar system is a 54- (I mean 48-) dimensional mathematical object; and many more if you include satellites and asteroids. An economy with a million different commodities, each having its own price, lives in a million-dimensional space. Electromagnetism, which requires only six extra numbers to characterize the local states of the electric and magnetic fields, is child's play by comparison. Examples like these abound. As science became interested in systems with large numbers of variables, it was forced to come to grips with extravagantly multidimensional spaces.

The formal mathematics of multidimensional spaces is purely alge-braic, based on "obvious" generalizations from low-dimensional spaces. For example, every point in the plane (a two-dimensional space) can be specified by two coordinates, and every point in three-dimensional space can be specified by three coordinates. It is a short step to define a point in four-dimensional space as a list of four coordinates, and more generally to define a point in n-dimensional space as a list of n coordinates. Then n-dimensional space itself (or n-space for short) is just the set of all such points.

Similar algebraic machinations let you work out the distance between any two points in n-space, the angle between any two lines, and so on. From there on out, it's a matter of imagination: most sensible geometric shapes in two or three dimensions have straightforward analogues in n dimensions, and the way to find them is to describe the familiar shapes using the algebra of coordinates and then extend that description to n coordinates.

To get a feel for n-space, we must somehow equip ourselves with n-dimensional spectacles. We can borrow a trick from the English clergyman and schoolmaster Edwin Abbott Abbott, who in 1884 wrote a short book called *Flatland.* It is about the adventures of A. Square, who lived in the two-dimensional space of a Euclidean plane. Abbott does not tell us what the initial "A" stands for: I am convinced it should be "Albert," for reasons explained in my sequel *Flatterland,* and I will make that assumption here. Albert Square, a sensible sort, did not believe in the absurd notion of the third dimension until, one fateful day, a sphere passed through his planar universe and flung him into realms he could never have imagined.

Flatland was a satirical look at Victorian society embedded in a parable about the fourth dimension based on a transdimensional analogy. It's the analogy, not the satire, that concerns us here. Having successfully imag-ined yourself as a two-dimensional creature living in a plane, blissfully un-aware of the greater reality of 3-space, it is not so hard to imagine yourself as a three-dimensional creature living in 3-space, blissfully unaware of the greater reality of 4-space. Suppose Albert Square, sitting happily in Flat-land, wants to "visualize" a solid sphere. Abbott achieves this by making such a sphere pass through the plane of Flatland, moving perpendicular

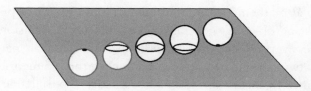

The sphere encounters Flatland.

to it so that Albert sees its cross-sections. First he sees a point, which grows to a circular disk. The disk expands until he is seeing the equator of the sphere, after which it shrinks again to a point and then vanishes.

Actually, Albert sees these disks edge-on, as line segments with graded shading, but his visual sense interprets this image as a disk, just as our stereo vision interprets a flat image as being solid.

By analogy, we can "see" a "hypersphere," the four-dimensional analogue of a solid sphere, as a point that grows to form a sphere, expands until we see the "equator," and then shrinks back to a point before disappearing.

Could space *really* have more than three dimensions? Not fancy mathematical fictions corresponding to nonspatial variables, but *real physical space?* After all, how can you fit the fourth dimension in? Everything's filled up already.

If you think that, you didn't listen to Albert Square, who would have argued exactly the same way about the plane. Ignoring our parochial prejudices, it seems that in principle, space might have been 4-dimensional, or million-dimensional, or whatever. Everyday observation, however, informs us that in our particular universe the good Lord settled on three dimensions for space, plus one for time.

Or did He? Whatever physics teaches us, one lesson is to be wary of everyday observation. A chair feels solid, but it's mostly empty space. Space looks flat, but according to relativity it's curved. Quantum physicists think that on very small scales space is a kind of quantum foam, mostly holes. And devotees of the "many worlds" interpretation of quantum uncertainty believe our universe is one of an infinite variety of coexisting

The hypersphere encounters Spaceland.

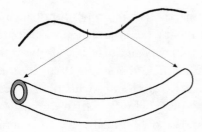

From a distance (above), a hosepipe looks one-dimensional. From close up (below) it has two additional dimensions.

universes, and that we occupy just a wafer-thin slice of a vast multiverse. If common sense can mislead us about those things, maybe it's wrong about the dimensionality of space or time.

Kaluza had a simple explanation for the extra dimension his theory assigned to space-time. The traditional dimensions point along straight lines, long enough to observe, billions of light years long, in fact. The new dimension, Kaluza suggested, is very different: it curls up tightly into a circle much smaller than an atom. The ripples that constitute light waves can move round the circle, because they, too, are much smaller than atoms, but matter cannot move in that direction because there isn't enough room.

This isn't such a silly idea. If you look at a hosepipe from a distance, the pipe looks like a curve, which is one-dimensional. Only from close up does it become clear that the pipe is really three-dimensional, with a small two-dimensional cross-section. This hidden structure in new dimensions explains something that you *can* observe from a distance: how the hose is able to carry water. The cross-section just needs to have the right shape, with a central hole. Now imagine that the thickness of the hose is less than the size of an atom. You would have to look extraordinarily closely to notice the extra dimensions. The incredibly thin hosepipe would no longer be able to carry water, but anything sufficiently small could still travel along it.

So it may be possible to perceive the effect of extra dimensions without perceiving the dimensions themselves. That means that hidden dimensions of space-time are an entirely scientific suggestion: their presence can in principle be tested—but by inference, rather than by a direct use of the senses. Most scientific tests work by inference—if you could see the cause of some phenomenon directly, you wouldn't need theories or experiments. No one has ever seen an electromagnetic field, for example. They have seen sparks and watched compass needles swing to point north, and (if they are scientists) they have inferred that a field must be responsible.

Kaluza's theory gained a certain popularity because it was the only known idea that held out hope of a unified field theory. In 1926, another mathematician, Oskar Klein, improved Kaluza's theory with the suggestion that quantum mechanics might explain why that fifth dimension curled up so tightly. In fact, its size should be of a similar order of magnitude to Planck's constant: the "Planck length" of 10^{-35} meters.

For a while, physicists were attracted to Kaluza–Klein theory, as it became known. But the impossibility of *directly* demonstrating the presence of that extra dimension preyed on their minds. By definition, Kaluza–Klein theory was consistent with every known phenomenon in gravitation and electromagnetism. You could never disprove it with standard experiments. But it didn't really add anything; it didn't predict anything new that could be tested. The same problem bedevils many attempts to unify existing laws. What's testable is already known, and what's new isn't testable. The initial enthusiasm began to wane.

The deathblow for Kaluza–Klein theory—not whether it was right, but whether it was worth spending precious research time on—was the explosive growth of a much sexier theory, one in which you really could make new predictions and do experiments to test them. This was quantum theory, then in its first flush of youth.

By the 1960s, though, quantum mechanics was running out of steam. Early progress had given way to deep puzzles and inexplicable observations. Quantum theory's success was undeniable, and it would shortly lead to the "standard model" of fundamental particles. But it was becoming ever more difficult to find new questions that had any chance of being answered. Genuinely novel ideas were too hard to test; testable ideas were mere extensions of existing ones.

One very elegant underlying principle had emerged from all the research: the key to the structure of matter on very tiny scales is *symmetry.* But the important symmetries for fundamental particles are not the rigid motions of Euclidean space, not even the Lorentz transformations of relativistic spacetime. They include "gauge symmetries" and "supersymmetries." And there are other kinds of symmetry, too, more like the symmetries studied by Galois, which act by permuting a discrete set of objects.

How can there be different kinds of symmetry?

Symmetries always form a group, but there are many different ways in which a group can *act*. It might act by rigid motions such as rotations, by permuting components, or by reversing the flow of time. Particle physics led to the discovery of a new way for symmetries to act, called gauge symmetries. The term is a historical accident, and a better name would be *local* symmetries.

Suppose that you are traveling in another country—let us call it Duplicatia—and you need money. The Duplicatian currency is the pfunnig, and the exchange rate is two pfunnigs to the dollar. You find this confusing until you notice a very simple and obvious rule for translating dollar transactions into pfunnigs. Namely, everything costs twice as many pfunnigs as you would expect to pay in dollars.

This is a kind of symmetry. The "laws" of commercial transactions are unchanged if you double all the numbers. To compensate for the numerical difference, though, you have to pay in pfunnigs, not in dollars. This "invariance under change of monetary scale" is a *global* symmetry of the rules for commercial transactions. If you make the same change throughout, the rules are invariant.

But now . . . Just across the border, in neighboring Triplicatia, the local currency is the boodle, and these are valued at three to the dollar. When you take a day trip to Triplicatia, the corresponding symmetry requires all sums to be multiplied by *three*. But again the laws of commerce remain invariant.

Now we have a "symmetry" that differs from one place to another. In Duplicatia, it is multiplication by two; in Triplicatia, by three. You would not be surprised to find that on visiting Quintuplicatia the corresponding multiple is five. All of these symmetry operations can be applied simultaneously, but each is valid only in the corresponding country. The laws of commerce are still invariant, but only if you interpret the numbers according to the correct local currency.

This local rescaling of currency transactions is a gauge symmetry of the laws of commerce. In principle, the exchange rate could be different at every point of space and time, but the laws would still be invariant provided you interpreted all transactions in terms of the local value of the "currency field."

❊

Quantum electrodynamics combines special relativity and electromagnetism. It was the first physical unification since Maxwell's, and it is based on a gauge symmetry of the electromagnetic field.

We have seen that electromagnetism is symmetric under the Lorentz group of special relativity. This group consists of global space-time symmetries; that is, its transformations must be applied consistently throughout the entire universe if we want to preserve Maxwell's equations. However, Maxwellian electromagnetism also has a gauge symmetry, which is vital to quantum electrodynamics. This symmetry is a *change of phase* in light.

Any wave consists of regular wobbles. The maximum size of the wobble is the amplitude of the wave. The time at which the wave hits that maximum is called its *phase*; the phase tells you when and where the peak value occurs. What really matters is not the absolute phase of any given wave but the difference in phases between two distinct waves. For example, if the phase difference of two otherwise identical waves is half the period (the time between maximum heights), then one wave hits its maximum exactly out of step with the other one, so the peaks of one coincide with the troughs of the other.

When you walk along the street, you left foot is half a period out of phase with your right foot. When an elephant walks along the street, successive feet hit the ground at phases 0, ¼, ½, and ¾ of the full period; first the left rear, then the left front, then the right rear, then the right front. You can appreciate that if we started counting from 0 at a different foot, we would get different numbers—but the phase *differences* would still be 0, ¼, ½, and ¾. So relative phases are well defined and physically meaningful.

Suppose a beam of light is passing through some complicated system of lenses and mirrors. The way it behaves turns out not to be sensitive to the overall phase. A change of phase is equivalent to a small time delay in making observations, or a resetting of the observer's clock. It does not af-

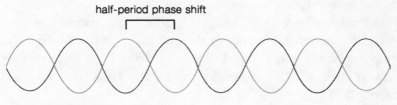

Effect of a phase shift on a wave.

fect the geometry of the system or the path of the light. Even if two light waves overlap, nothing changes, provided both waves have their phases shifted by the same amount.

So far, "change phase" is a global symmetry. But if an alien experimentalist somewhere in the Andromeda Galaxy changed the phase of light in one of its experiments, we would not expect to notice any effect inside a terrestrial laboratory. So the phase of light can be changed at will at all locations in space and time, and the laws of physics should remain invariant. The possibility of changing the phase arbitrarily at each point of spacetime, with no global constraint to make the same change everywhere, is a gauge symmetry of Maxwell's equations, and it carries over into the quantum version of those equations, quantum electrodynamics.

A phase shift of a full vibrational period is the same as no phase shift at all, and this implies that in the abstract, changing phase is a *rotation*. So the symmetry group involved here—the "gauge group"—is SO(2), the rotation group in two dimensions. However, physicists like their quantum coordinate transformations to be "unitary"—defined by complex numbers, not real ones. Fortunately, SO(2) has another incarnation as the unitary group U(1)—rotations in the *complex* plane.

In short: quantum electrodynamics has U(1) gauge symmetry.

Gauge symmetries were the clue to the next two unifications of physics, the electroweak theory and quantum chromodynamics. Together these constitute the "standard model," the currently accepted theory of all fundamental particles. Before we can see how this goes, we must explain exactly what is being unified: not theories but *forces*.

Today's physics recognizes four distinct kinds of force in nature: gravity, electromagnetism, the weak nuclear force, and the strong nuclear force. They have very different characteristics: they operate on different scales of space and time, some cause particles to attract each other, some cause them to repel each other, some do both depending on the particles, and some do both depending on how far apart the particles are.

At first sight, each force bears little resemblance to the others. But beneath the surface there are signs that these differences are less important than they seem. Physicists have teased out evidence of a deeper unity, suggesting that all four forces have a common explanation.

We feel the consequences of gravity all the time. When we drop a plate and it shatters on the kitchen floor, we see gravity pulling it towards the Earth's center and the floor getting in the way. The plastic pigs on the freezer door (well, that's what you will find in *our* house) remain in place thanks to the magnetic force, which Maxwell showed was merely one aspect of the unified electromagnetic force. The electrical aspect runs the freezer. Less obviously, the shattering plate also reveals the consequences of the electromagnetic force, because this is the main force acting in chemical bonds to hold bulk matter together. When the stress on the plate becomes too great for the electromagnetic force to hold its molecules together, it breaks.

The two remaining forces, which act on the level of the atomic nucleus, are not so readily apparent; but without them there would not be any matter at all, because they hold atoms together. They are why the plate, pigs, freezer, floor, and kitchen exist.

Other types of force could in principle give rise to other types of universe, and our ignorance of such possibilities is almost total. It is often claimed that without the particular forces we have, life would be impossible, proving that our universe is amazingly finely tuned to make life possible. This argument is bogus, a wild exaggeration based on too limited a view of what constitutes life. Life *like ours* would be impossible—but it is the height of arrogance to assume that our kind of life is the only kind of organized complexity that could exist. The fallacy here is to confuse *sufficient* conditions for life (those aspects of our universe on which our kind of life depends) with necessary ones.

The first of the four forces to be formulated scientifically was gravity. As Newton observed, this is an attractive force: any two particles in the universe, he said, attract each other gravitationally. The force of gravity is long-range: it falls off fairly slowly with distance. On the other hand, the gravitational force is much weaker than the other three: a tiny magnet can attach a plastic pig firmly to the fridge, even though the entire Earth is trying to pull it off through the force of gravity.

The next fundamental force to be isolated was electromagnetism, under whose influence particles may either attract or repel each other. What distinguishes the two is whether the particles have the same electric charge or the same magnetic polarity. If they do, the force is repulsive; if not, it is attractive. Again, this force is long-range.

The nucleus of an atom is assembled from smaller particles—protons and neutrons. Neutrons, as the name suggests, have no electric charge, but all protons have positive charge. The electromagnetic repulsion among protons should cause the nucleus to explode. What holds it together? Gravity is too weak—think of the plastic pigs. There must be some other force—which physicists labeled the strong nuclear force.

But if the strong force can overcome electric repulsion, why don't all of the protons in the universe get sucked together into one gigantic atomic nucleus? Clearly, the effect of the strong force must fall off rapidly at distances greater than the size of the nucleus. So the strong force is short-range.

The strong force does not explain the phenomenon of radioactive decay, in which atoms of certain elements "spit out" particles and radiation, and change to different elements. Uranium, for example, is radioactive and eventually turns into lead. So there must be yet another subatomic force. This is the weak force, and it is even shorter-range than the strong force: it acts only at a distance one-thousandth the size of a proton.

Physics was a lot easier when the only building blocks of matter were protons, neutrons, and electrons. These "elementary particles" were the components of the atom—which, it transpired, *did* split, even though the name means "indivisible." In Niels Bohr's early model, the atom was visualized as a tight collection of protons and neutrons orbited by much lighter, distant electrons. The proton carried a fixed positive electric charge, the electron carried the same amount of charge but negative, and the neutron was electrically neutral.

Later, as quantum theory developed, this solar-system image was replaced by a subtler one. The electrons didn't orbit the nucleus as well-defined particles but kind of smeared themselves around it in interestingly shaped clouds. These clouds were best interpreted as clouds of probability. If you looked for an electron, you were more likely to find it in the cloud's denser regions and less likely to find it in the sparse regions.

Physicists invented new ways to probe the atom, break it into pieces, and probe the inner structure of those pieces. The main method, still in use, is to hit it with another atom or particle and watch what flies off. Over time—the story is too complicated to tell in detail—more and more different kinds of particle were found. There was the neutrino, which could pass through a million miles of lead unhindered and was therefore rather hard to detect. There was the positron, like an electron but with the opposite electrical charge, predicted by Dirac's matter/antimatter symmetry.

As the number of "elementary" particles grew to more than sixty, physicists began to seek deeper ordering principles. There were too many building blocks for them to be fundamental. Each type of particle could be characterized by a series of properties: mass, charge, something called "spin" because the particles behaved as though they were spinning around some axis (except that this was an outmoded image and whatever spin was, it wasn't really that). The particles did not spin in space, like the Earth or a spinning top. They "spun"—whatever that meant—in more exotic dimensions.

Like everything in the quantum world, most of these features came in integer multiples of basic, very tiny amounts—quanta. All electrical charges were integer multiples of the charge on a proton. All spins were integer multiples of the spin of an electron. It was not clear whether mass was similarly quantized; the masses of the fundamental particles were a structureless mess.

Some family resemblances emerged. An important distinction had to be made between particles whose spin was an *odd* integer multiple of the spin of the electron, and those whose spin was an *even* integer multiple. The reason was based on symmetry properties; the spins (in those exotic dimensions) did different things if you made the particles rotate in space. Somehow the exotic dimensions of spin and the prosaic dimensions of space were related.

The odd particles were named fermions and the even ones bosons, after two giants of particle physics, Enrico Fermi and Satyendranath Bose. For reasons that once seemed sensible, the electron spin is defined to have value ½. So bosons have integer spin (even multiples of ½ are integers) and fermions have spins ½, ³⁄₂, ⁵⁄₂, and so on, along with their negatives – ½, – ³⁄₂, – ⁵⁄₂. Fermions obey the Pauli exclusion principle, which says that in any prescribed quantum system, two distinct particles cannot be in the same state at the same time. Bosons do not obey the Pauli principle.

Fermions include all of the familiar particles: the proton, neutron, and electron are all fermions. So are more esoteric particles like the muon, tauon, lambda, sigma, xi, and omega, all names derived from the Greek alphabet. So are three types of neutrino, associated with the electron, muon, and tauon.

Bosons are more mysterious, with names like pion, kaon, and eta.

The particle physicists knew that all of these particles existed, and they could measure their physical properties. The problem was making sense of this apparent mishmash. Was the universe built from whatever happened to be to hand? Or was there a hidden plan?

The upshot of these deliberations was that many supposedly elementary particles were in fact composite. They were all made from quarks. Quarks (the name comes from *Finnegans Wake*) come in six distinct flavors, arbitrarily named: up, down, strange, charm, top, and bottom. They are all fermions, with spin ½. Each has an associated antiquark.

There are two ways to combine quarks. One is to use three ordinary quarks, in which case you end up with a fermion. The proton consists of two up quarks plus one down quark. The neutron is two down and one up. A bizarre particle called the omega-minus is made from three strange quarks. The other is to use a quark and an antiquark, which yield a boson. They don't annihilate each other because they are kept apart by nuclear forces.

For the electrical charges to work out correctly, the charges on quarks cannot be integers. Some have charge ⅓, some ⅔. Quarks come in three distinct "colors." That makes 18 types of quark, plus 18 antiquarks. Oh, yes, there's more. We have to add some more particles to "carry" the weak nuclear force, which holds the quarks together. The resulting theory, which has great mathematical elegance despite the proliferation of particles, is known as quantum chromodynamics.

Quantum theory explains all physical forces in terms of exchanges of particles. Just as the tennis ball holds the two players together at opposite ends of the court as long as the game continues, so various particles carry the electromagnetic, strong, and weak forces. The electromagnetic force is carried by photons. The strong force is carried by gluons, the weak force by intermediate vector bosons, otherwise known as "weakons." (Don't blame me—I didn't invent these names, which are mostly historical accidents.) Finally, it is widely conjectured that gravity must be carried by a hypothetical particle called the *graviton*. No one has yet observed a graviton.

The large-scale effect of all these carrier particles is to fill the universe with "fields." Gravitational interactions create a gravitational field, electromagnetic ones create an electromagnetic field, and the two nuclear forces together create something called a Yang–Mills field, after the physicists Chen Ning Yang and Robert Mills.

We can summarize the main characteristics of the fundamental forces in a kind of physicist's shopping list:

- *Gravity:* Strength 6×10^{-39}, range infinite, carried by gravitons (not observed, should have mass 0, spin 2), forms the gravitational field.

- *Electromagnetism:* Strength 10^{-2}, range infinite, carried by photons (mass 0, spin 1), forms the electromagnetic field.

- *Strong force:* Strength 1, range 10^{-15} meters, carried by gluons (mass 0, spin 1), forms one component of the Yang–Mills field.

- *Weak force:* Strength 10^{-6}, range 10^{-18} meters, carried by weakons (large mass, spin 1), forms the other component of the Yang–Mills field.

You may feel that 36 fundamental particles, plus assorted gluons, is not a big improvement on sixty or more, but quarks form a highly structured family with a huge amount of symmetry. They are all variations on the same theme—unlike the wild zoo of particles that physicists had to deal with before quarks were discovered.

The description of fundamental particles in terms of quarks and gluons is known as the standard model. It fits experimental data extremely well. Some of the masses of some of the particles have to be adjusted to fit observations, but once you've done that, all the other masses slot neatly into place. The logic is not circular.

Quarks are bound together very tightly, and you never see an isolated quark. All you can observe are the combinations of twos and threes. Nevertheless, particle physicists have confirmed the existence of quarks indirectly. They're not just a clever numerological variation on the zoo. And to those who believe that the universe is at heart beautiful, the symmetry properties of quarks clinch it.

According to quantum chromodynamics, a proton is made from three quarks—two up, one down. If you took the quarks out of a proton, shuffled them, and put them back, you would still have a proton. So the laws for protons ought to be symmetric under permutations of their constituent quarks. More interestingly, the laws also turn out to be symmetric under changes to the *type* of quark. You could turn an up quark into a down one, say, and the laws would still work.

This implies that the actual symmetry group here is not just the group of six permutations of three quarks, but a closely related *continuous* group, SU(3), one of the simple groups on Killing's list. Transformations in SU(3) leave the equations for laws of nature unchanged, but they can change the *solutions* to those equations. Using SU(3) you can "rotate" a proton into a neutron, for instance. All you have to do is turn all of its quarks upside down, so that two up and one down become two down and one up. The world of fermions has SU(3) symmetry, and the symmetries act by changing one fermion into another.

Two further symmetry groups contribute to the standard model. The gauge symmetries of the weak force, SU(2), can change an electron into a neutrino. SU(2) is another group on Killing's list. And our old friend the electromagnetic field has U(1) symmetry—not the Lorentz symmetries of Maxwell's equations, but the gauge (i.e., local) symmetry of phase changes. This group just misses Killing's list because it is not SU(1), but it is morally on the list, since it's a very close relative.

The electroweak theory unified electromagnetism and the weak force by combining their gauge groups. The standard model incorporates the strong force as well, providing a single theory for all fundamental particles. It does this in a very direct manner: it just lumps all three gauge groups together as SU(3) × SU(2) × U(1). This construction is simple and straightforward but not terribly elegant, and it makes the standard model resemble something built out of chewing gum and string.

Suppose you own a golf ball, a button, and a toothpick. The golf ball has spherical symmetry SO(3), the button has circular symmetry SO(2), and the toothpick has (say) just a single reflectional symmetry O(1). Can you find some unified object that has all three types of symmetry? Yes, you can: put all three into a paper bag. Now you can apply SO(3) to the contents of the bag by rotating the golf ball, SO(2) by rotating the button, and O(1) by flipping the toothpick. The symmetry group of the bag's contents is SO(3) × SO(2) × O(1). This is how the standard model combines symmetries, but instead of using rotations it uses the "unitary transformations" of quantum mechanics. And it suffers from the same defect: it just lumps three systems together and combines their symmetries in an obvious, and rather trivial, way.

A much more interesting way to combine the three symmetry groups would be to build something that contained the same objects but was more elegant than a paper bag. Maybe you could balance the toothpick on the golf ball and stick a button on the end of it. You could even have a whole system of toothpicks, like the spokes of a wheel; put the button at the hub, spin the wheel on top of the golf ball. If you were really clever, maybe the combined object would have lots and lots of symmetry, say the group K(9). (There is no such group. I made it up for the sake of this discussion.) The separate symmetry groups SO(3), SO(2), and O(1) might with luck be subgroups of K(9). That would be a far more impressive way to unify the golf ball, button, and toothpick.

Physicists felt the same way about the standard model, and they wanted K(9) to be something on Killing's list or very close to it, because Killing's groups are the fundamental building blocks of symmetry. So they invented a whole series of Grand Unified Theories, or GUTs, based on groups like SU(5), O(10), and Killing's mysterious exceptional group E_6.

The GUTs seemed to suffer from the same defect as Kaluza–Klein theory—a lack of testable predictions. But then a really interesting prediction appeared. It was certainly new, so new that it seemed unlikely to be true, but it was testable. All GUTs predict that the proton can be "rotated" into an electron or a neutrino. So protons are *unstable,* and in the long run all matter in the universe should decay into radiation. The calculations said that on average, the life of a proton should be around 10^{29} years, much longer than the age of the universe. But individual protons would sometimes decay much sooner, and if you had enough protons, you might spot one decaying.

A big tank of water has more than enough protons for a few to decay each year. By the end of the 1980s there were six experiments running, all trying to spot a decaying proton. The biggest tank contained over 3000 tons of extremely pure water. No one saw a proton decay. Not one. Which meant that the average lifetime is at least 10^{32} years. Protons live at least a thousand times longer than GUTs predict. GUTs just don't hack it. In retrospect, it would have been a bit embarrassing if proton decay had been detected, because something very important is missing from GUTs: gravity.

✳

Any Theory of Everything has to explain why there are four fundamental forces, and why they take the strange forms that they do. This is a bit like trying to find a family resemblance between an elephant, a wombat, a swan, and a gnat.

It would be much easier to organize the four forces if they could all be shown to be different aspects of a single force. In biology, this has been achieved: elephants, wombats, swans, and gnats are all members of the Tree of Life, united by their DNA, distinguished by a lengthy series of historical changes to DNA. All four evolved, step by step, from a common ancestor, which lived a billion or two billion years ago.

The common ancestor of elephants and wombats is more recent than that of, say, elephants and swans. So this divergence constitutes the most recent branching of the tree of these four species. Before that, the common ancestor of elephants and wombats split off from some ancestor of the swan. Earlier still, the common ancestor of these three species split from that of the gnat.

Speciation can be viewed as a kind of symmetry-breaking. A single species is (approximately) symmetric under any permutation of its organisms; every wombat resembles every other wombat. When there are two distinct species—wombats and elephants—you can permute the wombats among themselves, and permute the elephants, but you can't change an elephant into a wombat without someone noticing.

The physicists' explanation of the underlying unity of the four forces is similar. The role of DNA, however, is played by the *temperature* of the universe—that is, its energy level. Although the underlying laws of nature are the same at all times, they lead to different behavior at different energies—

gnat swan wombat elephant

present

past

How four species diverge as time passes.

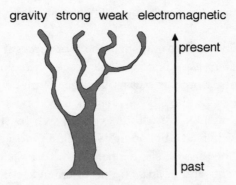

gravity strong weak electromagnetic

present

past

How the four fundamental forces diverge as time passes.

just as the same laws cause water to be solid at low temperatures, liquid at medium ones, and a gas at high ones. At very high temperatures, the water molecules break up to form a *plasma,* composed of separate particles. At higher temperatures still, the particles themselves break to form a quark–gluon plasma.

When the universe first came into being at the Big Bang, 13 billion years ago, it was enormously hot. At first, all four forces acted in exactly the same way. But as the universe cooled, its symmetry broke, and the forces split into individuals with distinguishable characteristics. Our present universe, with its four forces, is an imperfect shadow of that elegant original—the result of three broken symmetries.

14

THE POLITICAL JOURNALIST

n June 1972, during the run-up to the U.S. presidential election, a security guard at the Watergate complex noticed that a door had been taped open. He removed the tape, thinking it must have been left accidentally by workmen, but when he returned, someone had put it back. His suspicions aroused, the guard informed the police, who caught five men breaking into the offices of the Democratic Party's national committee. It turned out that the men were associated with President Nixon's reelection committee.

The discovery had little effect on the election itself; Nixon won in a landslide. But the story wouldn't go away, and slowly the tentacles of the Watergate affair reached higher and higher in the Nixon administration. Two reporters from the *Washington Post,* Bob Woodward and Carl Bernstein, pursued the story with dogged persistence, assisted by the clandestine revelations of "Deep Throat." No one knew who he was, but it was clear that he had to be a very senior official. In 2005, Deep Throat was revealed as Mark Felt, the second-in-command in the Federal Bureau of Investigation.

The information that Deep Throat leaked to the press was dynamite. By April 1974, Nixon had been forced to ask for the resignations of two senior aides. Then it turned out that the president had bugged his own office, and there were recorded tapes of sensitive conversations. After a legal battle to secure access to the tapes, gaps were found in some of the recordings, apparently the result of deliberate erasure.

The attempt to cover up the relation between the burglary and the White House was almost universally perceived as a worse crime than the burglary itself. The House of Representatives began a formal process that could lead to the president being impeached—tried for "high crimes and misdemeanors" before the U.S. Senate, and if found guilty, removed from office. When impeachment and conviction became inevitable, Nixon resigned.

Nixon's opponent in the election was Senator George McGovern. Announcing his candidacy for the Democratic nomination in Sioux Falls, South Dakota, McGovern made some prophetic remarks:

> Today, our citizens no longer feel that they can shape their own lives in concert with their fellow citizens. Beyond that is the loss of confidence in the truthfulness and common sense of our leaders. The most painful new phrase in the American political vocabulary is "credibility gap"—the gap between rhetoric and reality. Put bluntly, it means that people no longer believe what their leaders tell them.

Among the minor figures in McGovern's campaign was a would-be political journalist whose career would probably have taken off had McGovern been elected. In that variant of history, politics might have been richer, but fundamental physics and advanced mathematics would have been much the poorer. In the year 2004 of the history that actually happened, the journalist was listed by *Time* magazine as one of the year's one hundred most influential people—but not for his journalism.

Instead, he was listed for his groundbreaking contributions to mathematical physics. He is responsible for some of the most original mathematics in the world—for which he won the Fields Medal, the top honor in mathematics, comparable in prestige to a Nobel Prize—but he is not a mathematician. He is one of the world's leading theoretical physicists, and was awarded the National Medal of Science, but his first degree was in history. And he is the prime mover, though not quite the original creator, of the current front-runner in the effort to unify the whole of physics. He is the Charles Simonyi professor of mathematical physics at the Institute for Advanced Study in Princeton, where Einstein used to work, and his name is Edward Witten.

Like the great German quantum theorists but unlike poor Dirac, Witten grew up in an intellectual environment. His father, Louis Witten, is also a physicist, working on general relativity and gravitation. Edward was born in Baltimore, Maryland, and studied for his first degree at Brandeis University. After Nixon's reelection, he went back to academic life, taking a PhD at Princeton University, and embarked on a career of research and teaching at various American universities. In 1987 he was appointed to the Institute for Advanced Study, where all academic positions focus purely on research, and this is where he currently works.

Witten started research in quantum field theory, the first fruits of efforts to reconcile quantum theory with relativity. Here relativistic effects of motion are taken into account, but only in flat space-time. (Gravity, which requires curved space-time, is not considered.) In 1998, in a Gibbs lecture, Witten said that quantum field theory "encompasses most of what we know of the laws of physics, except gravity. In its seventy years there have been many milestones, ranging from the theory of 'antimatter' . . . to a more precise description of atoms . . . to the 'standard model of particle physics.'" He pointed out that having been developed largely by physicists, much of it lacked mathematical rigor and so had had little impact on mathematics as such.

The time was ripe, Witten said, to remedy that shortcoming. Several major areas of pure mathematics were effectively quantum field theory in disguise. Witten's own contribution, the discovery and analysis of "topological quantum field theories," had a direct interpretation in terms of concepts that various pure mathematicians had invented in quite different settings. These included the English mathematician Simon Donaldson's epic discovery that four-dimensional space is unique in supporting many different "differentiable structures"—coordinate systems in which calculus can be carried out. Other aspects are a recent breakthrough in knot theory, known as the Jones polynomial, a phenomenon called "mirror symmetry" in multidimensional complex surfaces, and several areas of modern Lie theory.

Witten made a bold prediction: a major theme in twenty-first-century mathematics would be the integration of ideas from quantum field theory into the mathematical mainstream:

> One has here a vast mountain range, most of which is still covered with fog. Only the loftiest peaks, which reach above the clouds, are seen in the mathematical theories of today, and these splendid peaks are studied in isolation . . . Still lost in the mist is the body of the range, with its quantum field theory bedrock and the great bulk of the mathematical treasures.

Witten's Fields Medal celebrated his uncovering of a few of those hidden treasures. Among them was a new and improved proof of the "positive mass conjecture," to the effect that a gravitational system with positive local mass density must have positive total mass. It may sound obvious, but in the quantum world mass is a subtle concept. The proof of this long-sought result, published by Richard Schoen and Shing-Tung Yau in 1979, had earned Yau a Fields Medal in 1982. Witten's new, improved proof exploited "supersymmetry," the first application of that concept to a significant problem in mathematics.

We can understand supersymmetry in terms of an old puzzle, which asks for a cork that can fit into a bottle whose opening may be circular, square, or triangular. Amazingly, such shapes do exist, and the traditional answer is a cork with a circular base that tapers like a wedge. Viewed from below, it looks like a circle; viewed from the front it is a square; viewed from the side it is a triangle. A single shape can perform all three tasks because a three-dimensional object can have several different "shadows," or projections, in different directions.

Now, imagine a Flatlander living on the "floor" of my picture, able to observe the projection of the cork onto the floor but unaware of the other projections. One day he discovers to his amazement that the circular shape has somehow morphed into a square. How can that be? It's certainly not a symmetry.

Not in Flatland. But while the Flatlander's back was turned, someone living in three dimensions rotated the cork so that its projection onto the

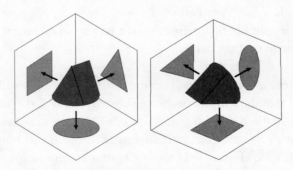

How supersymmetry works. Left: A cork to fit three shapes of hole. Right: Effect of rotating the cork.

floor changed to a square. And rotation *is* a symmetry transformation in three dimensions. So a symmetry in a higher dimension can sometimes explain a rather baffling transformation in a lower dimension.

Something very similar happens in supersymmetry, but instead of changing circles into squares, it changes fermions into bosons. This is amazing. It means that you can do calculations with fermions, hit everything with a supersymmetry operation, and deduce results for bosons with no extra effort. Or the other way round.

We expect this kind of thing to happen with genuine symmetries. If you stand in front of a mirror and juggle several balls, then whatever happens on your side of the mirror completely determines what happens on the other side. There, an image of you juggles images of the balls. If it takes 3.79 seconds to complete one sequence of juggles on the real side of the mirror, you know without doing the measurements that it will also take 3.79 seconds to complete the corresponding sequence of juggles on the other side. The two situations are related by a reflectional symmetry; whatever happens in one also happens, reflected, in the other.

Supersymmetries are not as obvious as this, but they have a similar effect. They let us deduce features of one type of particle from features of an entirely different type of particle. It is almost as if you could reach into some higher-dimensional region of the universe and twist a fermion into a boson. Particles come in supersymmetric pairs: an ordinary particle is matched with its twisted version, called a sparticle. Electrons are paired with selectrons, quarks with squarks. For historical reasons the photon is twinned not with the sphoton but with the photino. There is a kind of "shadow world" of sparticles that interacts only weakly with the ordinary world.

This idea makes for elegant mathematics, but the masses of these predicted shadow particles are too great for them to be observed in experiments. Supersymmetry is beautiful, but it may not be true. But even though direct confirmation is out of the question, indirect confirmation is still possible. Science mainly checks theories through their implications.

Witten pursued supersymmetry vigorously, and in 1984 he wrote an article titled "Supersymmetry and Morse theory." Morse theory is an area of topology, named for the pioneer Marston Morse, that relates the overall shape of a space to its peaks and valleys. Sir Michael Atiyah, probably Britain's most distinguished living mathematician, described Witten's paper as "obligatory reading for geometers interested in understanding modern quantum field theory. It also contains a brilliant proof of the classic Morse inequalities . . . The real aim of the paper is to prepare the ground for supersymmetric quantum field theory [in terms of] infinite-dimensional manifolds." Subsequently, Witten applied these techniques to other hot topics at the frontiers of topology and algebraic geometry.

It should be obvious that when I said Witten is not a mathematician, I did not mean he lacks mathematical ability—quite the reverse. Arguably no one on the planet has more mathematical ability. But in Witten's case it is complemented by an amazing physical intuition.

Unlike mathematicians, physicists are seldom shy about employing physical intuition to paper over any gaps in mathematical logic. Mathematicians have learned to regard leaps of faith with suspicion, however strong the supporting evidence may be: to them, proof is all. Witten is unusual in that he can relate his intuition to mathematics as mathematicians understand it. Atiyah puts it like this: "His ability to interpret physical ideas in mathematical form is quite unique. Time and again he has surprised the mathematical community by his brilliant application of physical insight leading to new and deep mathematical theorems."

But there is a flipside to this intuitive prowess. Many of Witten's most important ideas, being derived from physical principles or analogies, were arrived at without proofs, and some still lack proofs today. It's not that he can't *do* proofs—as his Fields Medal demonstrates—but he can make leaps of logic that lead to deep and correct mathematics without seeming to need proofs.

The big question is, does Witten's wonderfully elegant mathematics have anything to do with fundamental physics? Or has the search for beauty headed down a mathematical dead end, losing any connection with physical truth?

By 1980, physicists had unified three of the four forces of nature: electromagnetic, weak, and strong. But Grand Unified Theories said nothing about gravity. The force that we experience most directly in everyday life, which literally keeps our feet on the ground, was embarrassingly absent from the synthesis.

It was easy enough to write down combined theories of gravity and quantum theory that looked sensible. But whenever anyone tried to solve the resulting equations, they got nonsense. Typically, numbers that ought to represent reasonable physical quantities were infinite. An infinity in a physical theory is a sign that something is wrong. It was an infinity in the radiation law that inspired Planck to quantize light.

Some physicists became convinced that the main source of the infinities was the ingrained habit of treating particles as points. A point—location without size—is a mathematical fiction. Quantum particles were probabilistic fuzzed-out points, but that didn't cure the disease; something more drastic was needed. Even in the 1970s a few pioneers had begun to think that particles might more sensibly be modeled as tiny vibrating loops—"strings." In the 1980s, when supersymmetry got in on the act, these mutated into *superstrings*.

One could write an entire book about superstrings, and several people have, but we can manage with a rough hand-waving description. I want to focus on four features: the way relativistic and quantum pictures are combined, the need for extra dimensions, the interpretation of quantum states as vibrations in those extra dimensions, and the symmetries of the extra dimensions—or, more accurately, of various fields that live in them.

Our starting point is Einstein's idea of representing the trajectory of a particle in space-time as a curve, which he called its *world line*. Essentially, this is the curve that the particle traces in space-time as it moves. In relativity, world lines are smooth curves, because of the form of Einstein's field equations. They do not branch, because in relativity the future of any system is completely determined by its past, indeed by its present.

There is an analogous concept in quantum field theory called a Feynman diagram. Feynman diagrams depict the interaction of particles in a rather schematic space-time. For example, the left-hand picture (next page) is the Feynman diagram for an electron that emits a photon that is then captured by a second electron. It is traditional to use wiggly lines for photons.

The Feynman diagram is a bit like a relativistic world line, but it has sharp corners and it branches. In 1970, it occurred to Yoichiro Nambu that if the assumption that particles are points is replaced by the assumption that they are tiny loops, then Feynman diagrams can be converted into smooth surfaces—*worldsheets*—as in the right-hand picture. A worldsheet can be interpreted as a world line in a modified space-time, with an extra dimension for the loops to live in.

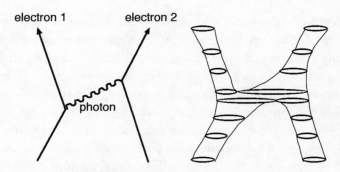

(Left) Feynman diagram for interacting particles. (Right) The corresponding worldsheet, sliced into strings.

The great thing about loops—aside from not being points—is that they can vibrate. Perhaps each vibrational pattern corresponded to a quantum state. That would explain why quantum states come in whole-number multiples of some basic quantity—for example spin, which is always an integer multiple of ½. The number of waves that fit into the loop has to be a whole number. In a violin string, these different patterns are the fundamental note and its higher harmonics. So quantum theory becomes a kind of music, played with superstrings instead of violin strings.

Nambu's idea did not come out of the blue. It had its roots in a remarkable formula derived by Gabriele Veneziano in 1968, which showed that apparently distinct Feynman diagrams represent the same physical process, and that any failure to take that into account leads to wrong answers in quantum field theory calculations. Nambu noticed that when the Feynman diagram is surrounded by tubes, different diagrams yield networks of tubes with the same topology. That is, these networks can be deformed into each other. So Veneziano's formula seemed to be related to the topological properties of the tubes.

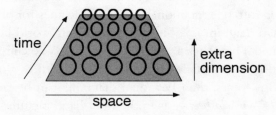

Strings poke out of ordinary space-time into new dimensions.

This, in turn, hinted that quantum particles, with their discrete quantum numbers like charge, might be topological features of a smooth spacetime. Mathematicians had already observed the tendency of basic topological properties—such as the number of holes in a surface—to be discrete. It all seemed to fit. But as always, the devil was in the detail, and the detail was devilish. String theory was the first attempt to get the detail in agreement with the real world.

String theory did not start out as a possible route to a Theory of Everything but as a proposal to explain the particles collectively known as hadrons. These include most of the common particles found in the atomic nucleus, such as the proton and neutron, together with a host of more exotic particles. However, the theory had a flaw: it predicted the existence of a particle with zero mass and spin 2, which had never been observed (and still hasn't). Additionally, it failed to predict any particles with spin ½—and rather a lot of hadrons, including the proton and neutron, have spin ½. It was like a midsummer weather forecast that predicts hailstones a foot across but has nothing to say about whether the temperature will be warm. Physicists were unimpressed. In 1974, when quantum chromodynamics came along and explained all known hadrons, and even successfully predicted a new one, the omega-minus, the fate of string theory seemed sealed.

At that point, however, John Schwarz and Joel Scherk noticed that string theory's unwanted zero-mass spin-2 particle might be the long-sought graviton, the hypothetical particle believed to carry the force of gravity. Might string theory be a quantum theory of *gravity,* rather than hadrons? If so, it would be an attractive contender for a Theory of Everything—well, for a Theory of Many Things, because there are many particles that are not hadrons.

At this point, supersymmetry came into play, because it converts fermions into bosons. Hadrons include particles of both kinds, but other particles, such as the electron, are not hadrons. If supersymmetry could be incorporated into string theory, then a number of new particles would automatically come within the theory's grasp—carried along by supersymmetric partners that were already part of the theory.

The combined theory, developed by Pierre Ramond, André Neveu, and Schwarz, was superstring theory. This theory did include spin-½ particles, and it eliminated a nasty feature of ordinary string theory, a particle that goes faster than light. The presence of such a particle in a theory is now seen as evidence that it is unstable, which rules it out.

From 1980 onward, Michael Green, a British theoretical physicist, worked out more and more of the mathematics of superstrings, using techniques from Lie group theory and topology, and it quickly became clear that whatever its physical credentials, superstring theory possessed extraordinary mathematical beauty. The physics remained obstinate: in 1983, Luis Alvarez-Gaume and Witten discovered a new snag with string theories, including superstrings and even good old quantum field theory. Namely, these theories normally possess *anomalies*. An anomaly occurs when the process of converting a classical system to its quantum analogue changes an important symmetry.

Green and Schwarz had discovered that very occasionally, the anomalies miraculously disappear, but only if space-time has 26 dimensions (in the first version of the theory, called bosonic string theory) or 10 dimensions (in later modifications). Why? In their calculations for bosonic string theory, the mathematical terms that would create an anomaly are multiplied by $d - 26$, where d is the dimension of space-time. So these terms vanish precisely when $d = 26$. Similarly, in the modified version, the factor becomes $d - 10$. Time always remains one-dimensional, but space somehow acquires an extra 6 or 22 dimensions. Schwarz put it this way:

> In 1984 Michael Green and I did a calculation for one of these superstring theories to see whether, in fact, this anomaly occurred or not. What we discovered was quite surprising to us. We found that, in general, there was indeed an anomaly that rendered the theory unsatisfactory. Now there was freedom to choose the particular symmetry structure that one used in defining a theory in the first place. In fact, there were an infinite number of possibilities for these symmetry structures. However, for just one of them the anomaly magically cancelled out of the formulae, whereas for all the others it didn't. So amid this infinity of possibilities, just one unique one was being picked out as being potentially consistent.

If you were prepared to ignore the weird numbers 10 or 26, this discovery was very exciting. It suggested that there might be a mathematical reason for space-time to have a particular number of dimensions. It was disappointing that the number was not four, but it was a start. Physicists had always wondered why space-time has the dimensions it does; now it looked as though there might be a better answer to that question than, "well, it could be anything, but in our universe it's four."

Perhaps other theories would lead to a four-dimensional space-time. It would have been ideal, but nothing along those lines seemed to work, and the funny dimensions refused to go away. So maybe they were *there*. This was an old idea of Kaluza's: space-time might have extra dimensions that we are unable to observe. If so, the strings would remain one-dimensional loops, but those loops would vibrate in an otherwise invisible higher-dimensional space. The quantum numbers associated with the particles, like charge or charm, would be determined by the form of the vibrations.

A basic question was, what do the hidden dimensions look like? What *shape* is space-time?

At first, physicists hoped the extra dimensions would form some simple shape like the 6-dimensional analogue of a torus. But in 1985, Philip Candelas, Gary Horowitz, Andrew Strominger, and Witten reasoned that the most suitable shape would be a so-called Calabi–Yau manifold. There are tens of thousands of these shapes; here is a typical one:

A Calabi–Yau manifold (schematic).
Credit: Andrew J. Hanson, Professor and Chair, Indiana University.

The great advantage of Calabi–Yau manifolds is that the supersymmetry of 10-dimensional space-time is inherited by the ordinary four-dimensional space-time that underlies it.

For the first time, the exceptional Lie groups were taking on a prominent role in frontier physics, and this trend accelerated. Around 1990, there seemed to be five possible types of superstring theory, all with space-time dimension equal to 10. The theories are called Type I, Types IIA and IIB, and "heterotic" types HO and HE. Interesting gauge symmetry groups turn up; for example, in Types I and HO we find SO(32), the rotation group in 32-dimensional space, and in Type HE the exceptional

Lie group E_8 turns up as $E_8 \times E_8$, two distinct copies acting in two different ways.

The exceptional group G_2 also makes an appearance in the latest twist to the story, which Witten calls M-theory. The "M," he says, stands for magic, mystery, or matrix. M-theory posits an 11-dimensional space-time, which unifies all five of the 10-dimensional string theories, in the sense that each can be obtained from M-theory by fixing some of its constants to particular values. In M-theory, Calabi–Yau manifolds are replaced by 7-dimensional spaces known as G_2-manifolds, because their symmetries are closely related to Killing's exceptional Lie group G_2.

At the moment there is a bit of a backlash against string theory; not on the grounds that it is known to be wrong, but on the grounds that it's not yet known to be right. Several prominent physicists, especially experimentalists, have never had much truck with superstrings anyway—mostly because it didn't give them anything to *do*. There were no new phenomena to observe, no new quantities to measure.

I'm not wedded to superstrings as the key to the universe, but I think this criticism is unfair. String theorists are being asked to prove their innocence, whereas normally it would be up to the critics to prove their guilt. It takes a lot of time and effort to develop radically new ways of thinking about the physical world, and string theory is technically very difficult. In principle, it *can* make new predictions about our world; the big problem is that doing the necessary sums is extraordinarily hard. The same complaint could have been made about quantum field theory 40 years ago, but eventually the sums got done, through a combination of better computers and better mathematics, and the agreement with experiment turned out to be better than we find anywhere else in science.

Moreover, much the same charge can be leveled at almost any hopeful Theory of Everything, and paradoxically, the better it is, the harder it will be to prove correct. The reason is inherent in the nature of a Theory of Everything. In order to be successful, it must agree with quantum theory whenever it is applied to any experiment whose results are consistent with quantum theory. It must also agree with relativity whenever it is applied to any experiment whose results are consistent with relativity. So the Theory of Everything is obliged to pass *every experimental test yet devised.* Asking for a new prediction that distinguishes the Theory of Everything from conventional physics is rather like asking for something that yields results identical to those predicted by theories describing all known physical phenomena, yet is different.

Of course, eventually string theory will have to make a new prediction, and be tested against observations, to make the transition from speculative theory to real physics. The need to agree with everything currently known does not rule out such predictions, it just explains why they don't come easily. Some tentative proposals for critical experiments already exist. For instance, recent observations of distant galaxies indicate that the universe is not only expanding, but expanding increasingly fast. Superstring theory offers a simple explanation—gravity is leaking away into those extra dimensions. However, there are other ways to explain this particular effect. What is clear is that if the theorists all stop investigating superstring physics, we will never have a chance to find out whether the theory is correct. It takes time and effort to come up with the crucial experiments, even if they exist.

I don't want to leave the impression that when it comes to unifying quantum theory with relativity, superstrings are the only game in town. There are many competing proposals—though they all suffer from the same lack of experimental support.

One idea, known as "noncommutative geometry," is the brainchild of the French mathematician Alain Connes. It rests on a new concept of the geometry of space-time. Most unifications start with the idea that space-time is some extension of Einstein's relativistic model, and try to make the fundamental particles of subatomic physics fit in somehow. Connes does the opposite. He starts from a mathematical structure known as a non-commutative space, which contains all of the symmetry groups that arise in the standard model, and then deduces features similar to relativity. The mathematics of such spaces traces back to Hamilton and his noncommutative quaternions, but is extensively generalized and modified. Once again, though, this alternative theory is firmly rooted in Lie group theory.

Another intriguing idea is "loop quantum gravity." In the 1980s, the physicist Abhay Ashtekar worked out how Einstein's equations would look in a quantum setting where space is "grainy." Lee Smolin and Carlo Ravelli developed his ideas, leading to a model of space that is rather like medieval chain mail—constructed from very tiny lumps about 10^{-35}

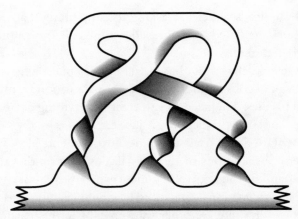

An electron represented as a braid.

meters across, joined by links. They noticed that the detailed structure of the chain mail can get very complex as the links become knotted or braided together. However, it was not clear what these possibilities meant.

In 2004, Sundance Bilson-Thompson discovered that some of these braids exactly reproduce the rules for combining quarks. The electric charge of the quark is reinterpreted in terms of the topology of the associated braid, and the combination rules follow from simple geometric operations with braids. This idea, still in its infancy, produces most of the particles observed in the standard model. It is the latest in a series of speculative proposals that matter—here realized as particles—might be a consequence of "singularities" in space, such as knots, localized waves, or more complicated structures where space ceases to be smooth and regular. If Bilson-Thompson is right, matter is just twisted space-time.

Mathematicians have been studying the topology of braids for many years and have long known that braids themselves form a group, the *braid group*. The operation of "multiplication" arises when two braids are joined end to end—much as we joined permutations end to end when discussing Ruffini's approach to the quintic. Yet again, physics is building on preexisting mathematical discoveries, mostly made "for their own sake" because they looked interesting. And yet again, a key ingredient is symmetry.

In the latest version of superstrings, the biggest problem is an embarrassment of riches. Instead of making no predictions, the theory makes too many. The "vacuum energy"—the energy content of empty space—can be almost anything, depending on how the strings wrap around inside the extra dimensions of space. The number of ways for this to happen is gigantic—around 10^{500}. Different choices yield different values for the vacuum energy.

As it happens, the observed value is very, very small, around 10^{-120}, but it is not zero.

According to the conventional "fine-tuning" story, this particular value is exactly right for life to exist. Anything larger than 10^{-118} makes local space-time explode; anything smaller than 10^{-120} and space-time contracts in a cosmic crunch and disappears. So the "window of opportunity" for life is very small. By a miracle, our universe sits neatly within it.

The "weak anthropic principle" points out that if our universe were not constituted the way it is, we wouldn't be here to notice, but that leaves open the question why there is a "here" for us to occupy. The "strong anthropic principle" says that we're here because the universe was designed specially for life to exist—which is mystical nonsense. No one actually knows what the possibilities would be if the vacuum energy were markedly different from what it is. We know a few things that would go wrong—but we have no idea what might go right instead. Most of the fine-tuning arguments are bogus.

In 2000, Raphael Bousso and Joseph Polchinski proposed a different answer, using string theory and taking advantage of those 10^{500} possible values for the vacuum energy. Although 10^{-120} is very small, the possible vacuum energy levels are spaced about 10^{-500} units apart, which is even smaller. So plenty of string theories give vacuum energies in the "right" range. The probability that a randomly chosen one will do that is still negligible, but Bousso and Polchinski pointed out that this is irrelevant. Eventually, the "right" vacuum energy will inevitably occur. The idea is that the universe explores all possible string theories, sticking with any given one until it causes that universe to come to bits, and then "tunneling" quantum-mechanically to some other string theory. If you wait long enough, then at some stage the universe acquires a vacuum energy that happens to be in the range suitable for life.

In 2006, Paul Steinhardt and Neil Turok proposed a variation on the "tunneling" theory: a cyclic universe that expands in a Big Bang and contracts in a Big Crunch, repeating this behavior every trillion years or so. In their model, the vacuum energy decreases in each successive cycle, so that eventually the universe has a very small, but nonzero, vacuum energy.

In either model, a universe whose vacuum energy is low enough will hang around for a very long time. Conditions are suitable for life to arise, and life has plenty of time to evolve intelligence, and to wonder why it's there.

15

A MUDDLE OF MATHEMATICIANS

gaggle of geese, a pride of lions, a charm of finches, an exaltation of skylarks . . . what is the collective noun for mathematicians? A magnificence of mathematicians? Too smug. A mystification of mathematicians? Too close to the mark. Having had many opportunities to observe the behavior of the mathematical species when congregated in large herds, I think that the most apt word is "muddle."

One such muddle invented one of the most bizarre structures in the entire subject, and discovered a hidden unity behind its puzzling facade. Their discoveries, mainly obtained by pottering around and seeing what turned up, are beginning to infiltrate theoretical physics, and they may just hold the key to some of the more curious features of superstrings.

The mathematics of superstrings is so new that most of it has not been invented yet. But ironically, mathematicians and physicist have just discovered that superstrings, at the frontiers of modern physics, seem to have a curious relationship to a bit of Victorian algebra so old-fashioned that it is seldom mentioned in university mathematics courses. This algebraic invention is known as the octonions, and it is the next structure in line after real numbers, complex numbers, and quaternions.

Octonions were discovered in 1843, published by someone else in 1845, and ever afterward credited to the wrong person—but that didn't matter, since nobody took any notice. By 1900 they had fallen into obscurity even in mathematics. They experienced a brief revival in 1925 when Wigner and von Neumann tried to make them the basis of quantum mechanics, but then fell back into obscurity when the attempt failed. In the 1980s they resurfaced as a potentially useful gadget in string theory. In

1999 they turned up as a crucial ingredient in 10- and 11-dimensional superstring theory.

Octonions tell us that there is something very strange about the number 8, and something even stranger about the physics of space, time, and matter. A Victorian whimsy has been reborn as the key to deep mysteries on the common frontier of mathematics and physics—especially the belief that space-time may have more dimensions than the traditional four, and that this is how gravity and quantum theory fit together.

The tale of the octonions lives in the heady realms of abstract algebra, and it is the topic of a beautiful mathematical survey published in 2001 by the American mathematician John Baez. I have drawn heavily on Baez's insights here. I'll do my best to convey the bizarre yet elegant wonders that inhabit this curious interface between mathematics and physics. As with the ghost of Hamlet's father, a disembodied voice beneath the stage, much of the mathematical action must happen out of sight of the audience. Bear with me, and don't worry too much about the odd piece of unexplained jargon. Sometimes we just need a convenient word to keep track of the main players.

A few reminders may help to set the scene. The step-by-step expansion of the number system has woven in and out of our tale of the quest for symmetry. The first step was the discovery (or invention) in the mid-sixteenth century of complex numbers, in which −1 has a square root. Until that time, mathematicians had thought that numbers were God-given, unique, and *done*. No one could contemplate inventing a new number. But around 1550, Cardano and Bombelli did just that, by writing down the square root of a negative number. It took about 400 years to sort out what the thing meant, but only 300 to convince mathematicians that it was too useful to be ignored.

By the 1800s, Cardano and Bombelli's baroque concoction had crystallized into a new kind of number, with a new symbol: *i*. Complex numbers may seem weird, but they turn out to be a marvelous tool for understanding mathematical physics. Problems of heat, light, sound, vibration, elasticity, gravity, magnetism, electricity, and fluid flow all succumb to the complex weaponry—but only for physics in two dimensions.

Our own universe, however, has three dimensions of space—or so we thought until recently. Since the two-dimensional system of complex numbers is so effective for two-dimensional physics, might there be an analogous three-dimensional number system that can be used for genuine physics? Hamilton spent years trying to find one, with absolutely no success. Then, on 16 October 1843, he had a flash of insight: *don't look in three dimensions, look in four,* and carved his equations for quaternions into the stonework of Brougham Bridge.

Hamilton had an old friend from college, John Graves, who was an algebra buff. It was probably Graves who got Hamilton excited about extensions of the number system in the first place. Hamilton wrote his buddy a long letter about quaternions the day after he had vandalized the bridge. Graves was initially perplexed, and wondered how legitimate it was to invent multiplication rules off the top of one's head. "I have not yet any clear views as to the extent to which we are at liberty arbitrarily to create imaginaries, and to endow them with supernatural properties," he wrote back. But he also saw the potential of the new idea and wondered how far it might be pushed: "If with your alchemy you can make three pounds of gold, why stop there?"

It was a good question, and Graves set about answering it. Within two months he wrote back to say that he had found an *eight*-dimensional number system. He called it the "octaves." Associated with it was a remarkable formula about sums of eight squares, to which I will shortly return. He tried to define a 16-dimensional number system, but met what he called an "unexpected hitch." Hamilton said that he would help to bring his friend's discovery to public attention, but he was too busy exploring quaternions to do so. Then he noticed a potential embarrassment: multiplication of octaves does not obey the associative law. That is, the two ways to form products of three octaves, $(ab)c$ and $a(bc)$, are usually *different*. After much soul-searching, Hamilton had been willing to dispense with the commutative law, but throwing away the associative law might be going too far.

Now Graves had some serious bad luck. Before he could publish, Cayley independently made the same discovery, and in 1845 he published it as an addendum to an otherwise awful paper on elliptic functions—so riddled with errors that it was removed from his collected works. Cayley called his system "octonions."

Graves was unhappy at being beaten to publication, and it so happened that a paper of his own was shortly due to be published in the journal where Cayley had announced his discovery. So Graves added a note to his paper, pointing out that he had come across the same idea two years before, and Hamilton backed him up by publishing a brief note confirming that his friend should be granted priority. Despite the record being set straight, the octonions quickly acquired the name "Cayley numbers," which is still widely used. Many mathematicians now use Cayley's terminology, calling the system "octonions," but give credit to Graves. It's a better name than "octaves" anyway, because it resembles "quaternions."

The algebra of octonions can be described in terms of a remarkable diagram known as the Fano plane. This is a finite geometry composed of seven points joined in threes by seven lines, and it looks like this:

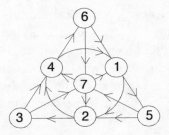

The Fano plane, a geometry with
seven points and seven lines.

One line has to be bent into a circle to draw it in the plane, but that doesn't matter. In this geometry, any two points are joined by a line, and any two lines meet at a point. There are no parallel lines. The Fano plane was invented for a totally different purpose, but it turned out to encapsulate the rules for multiplying octonions.

The octonions have eight units: the ordinary number 1, and seven others called e_1, e_2, e_3, e_4, e_5, e_6, and e_7. The square of any of these is -1. The diagram determines the multiplication rule for the units. Suppose you want to multiply e_3 by e_7. Look in the diagram for points 3 and 7 and find the line that joins them. On it, there is a third point, which in this case is 1. Following the arrows, you go from 3 to 7 to 1, so $e_3 e_7 = e_1$. If the ordering is back to front, throw in a minus sign: $e_7 e_3 = -e_1$. Do this for all possible pairs of units, and you know how to do arithmetic with octonions. (Addition and subtraction are always easy, and division is determined by multiplication.)

Graves and Cayley didn't know about this connection with finite geometry, so they had to write out a multiplication table for octonions. The Fano plane pattern was discovered later.

For many years, the octonions were merely a minor curiosity. Unlike quaternions, they had no geometrical interpretation and no application in science. Even within pure mathematics, nothing seemed to follow from them; no wonder they fell into obscurity. But all this would change with the realization that the octonions are the source of the most bizarre algebraic structures known to mathematics. They explain where Killing's five exceptional Lie groups—G_2, F_4, E_6, E_7, and E_8—really come from. And the group E_8, the largest of the exceptional Lie groups, shows up *twice* in the symmetry group that forms the basis of 10-dimensional string theory, which has unusually pleasant properties and is thought by many physicists to be the best candidate yet for a Theory of Everything.

If we agree with Dirac that the universe is rooted in mathematics, then we could say that a plausible Theory of Everything exists because E_8 exists, and E_8 exists because the octonions exist. Which opens up an intriguing philosophical possibility: the underlying structure of our universe, which we know to be very special, is singled out by its relationship to a unique mathematical object: the octonions.

Beauty is truth, truth beauty. The Pythagoreans and Platonists would have loved this evidence of the pivotal role of mathematical patterns in the structure of our world. The octonions have a haunting, surreal mathematical beauty, which Dirac would have seized upon as a reason why 10-dimensional string theory has to be true. Or, if unhappily proved false, is nevertheless more interesting than whatever *is* true. But we have learned that beautiful theories need not be true, and until the verdict on superstrings is in, this possibility must remain pure conjecture.

Whatever its importance in physics, the circle of ideas surrounding the octonions is pure gold for mathematics.

The connection between the octonions and the exceptional Lie groups is just one of many strange relationships between various generalizations of the quaternions and the frontiers of today's physics. I want to explore some of these connections in enough depth for you to appreciate how remarkable they are. And I'm going to start with some of the oldest exceptional structures in mathematics, formulas about sums of squares.

One such formula derives naturally from the complex numbers. Every complex number has a "norm," the square of its distance from the origin. The Pythagorean theorem implies that the norm of $x + iy$ is $x^2 + y^2$. The rules for multiplying complex numbers, as laid down by Wessel, Argand, Gauss, and Hamilton, tell us that the norm has a very pretty property. If you multiply two complex numbers together, then the norms multiply too. In symbols, $(x^2 + y^2)(u^2 + v^2) = (xv + yu)^2 + (xu - yv)^2$. A sum of two squares times a sum of two squares is always a sum of two squares. This fact was known to the Indian mathematician Brahmagupta around 650, and to Fibonacci in 1200.

The early number theorists were fascinated by sums of two squares, because they distinguished two different types of prime number. It is easy to prove that if an odd number is the sum of two squares, then it must be of the form $4k + 1$ for some integer k. The remaining odd numbers, which are of the form $4k + 3$, cannot be represented as the sum of two squares. However, it is *not* true that every number of the form $4k + 1$ is a sum of two squares, even if we allow one of the squares to be zero. The first exception is the number 21.

Fermat made a very beautiful discovery: these exceptions can never be prime. He proved that on the contrary, every *prime* number of the form $4k + 1$ is a sum of two squares. By applying the above formula for multiplying sums of two squares together, it then follows that an odd number is a sum of two squares if and only if every prime factor of the form $4k + 3$ occurs to an even power. For instance, $45 = 3^2 + 6^2$ is a sum of two squares. Its prime factorization is $3 \times 3 \times 5$, and the prime factor 3, which has the form $4k + 3$ with $k = 0$, occurs to the power two—an even number. The other factor, 5, occurs to an odd power, but that's a prime of the form $4k + 1$ (with $k = 1$), so it doesn't cause any trouble.

On the other hand, the exception 21 is equal to 3×7, which are both primes of the form $4k + 3$, and here each occurs to the power 1, which is odd—so that's why 21 doesn't work. Infinitely many other numbers don't work for the same reason.

Later, Lagrange used similar methods to prove that *every* positive whole number is a sum of four squares (zero permitted). His proof used a clever formula discovered by Euler in 1750. It is similar to the one above, but for sums of four squares. A sum of four squares times a sum of four squares is itself a sum of four squares. There can be no such formula for sums of three squares, because there exist pairs of numbers that are both sums of three squares but whose product is not. However, in 1818 Degen found a product formula for sums of *eight* squares. It is the same formula that Graves discovered using octonions. Poor Graves—his discovery of octonions, which was original, was credited to someone else; his other discovery, the eight-squares formula, turned out not to be original.

There is also a trivial product formula for sums of one square—that is, squares. It is $x^2y^2 = (xy)^2$. This formula does for real numbers what the two-squares formula does for complex numbers: it proves that the norm is "multiplicative"—the norm of a product is the product of the norms. Again, the norm is the square of the distance from the origin. The negative of any number has the same norm as its positive.

What of the four-squares formula? It does the same thing for quaternions. The four-dimensional analogue of the Pythagorean theorem (yes, there is such a thing) tells us that a general quaternion $x + iy + jz + kw$ has norm $x^2 + y^2 + z^2 + w^2$, a sum of four squares. The quaternionic norm is also multiplicative, and this explains Lagrange's four-squares formula.

You will probably be ahead of me by now. Degen's eight-squares formula has a similar interpretation for octonions. The octonion norm is multiplicative.

Something very curious is going on here. We have four types of ever-more-elaborate number system: the reals, complexes, quaternions, and octonions. These have dimensions 1, 2, 4, and 8. We have formulas that say that a sum of squares times a sum of squares is a sum of squares: these apply to 1, 2, 4, or 8 squares. The formulas are closely related to the number systems. More intriguing still is the pattern of the numbers.

1, 2, 4, 8—what comes next?

If the pattern continued, we would confidently expect to find an interesting 16-dimensional number system. Indeed, such a system can be constructed in a natural way, called the Cayley–Dickson process. If you apply that process to the reals, you get the complexes. Apply it to the complexes, you get the quaternions. Apply it to the quaternions, you get the octonions. And if you plow ahead and apply it to the octonions, you get the *sedenions,* a 16-dimensional number system, followed by algebras of dimension 32, 64, and so on, doubling at every step.

So there is a 16-squares formula, then?

No. The sedenion norm is not multiplicative. Product formulas for sums of squares exist *only* when the number of squares involved is 1, 2, 4, or 8. The law of small numbers strikes again: the apparent pattern of powers of two grinds to a halt.

Why? Basically, the Cayley–Dickson process slowly destroys laws of algebra. Every time you apply it, the resulting system is not quite as well behaved as the previous one. Step by step, law by law, the elegant real number system descends into anarchy. Let me explain in more detail.

The four number systems have other features in common aside from their norms. Their most striking feature, which qualifies them as generalizations of the real numbers, is that they are "division algebras." There are many algebraic systems in which notions of addition, subtraction, and multiplication are valid. But in these four systems, you can also divide. The existence of a multiplicative norm makes them "normed division algebras." For a while, Graves thought his method of going from 4 to 8 could be repeated, leading to normed division algebras with 16, 32, 64 dimensions, any power of two. But he hit a snag with the sedenions and began to doubt whether a 16-dimensional normed division algebra could exist. He was right: we now know that there exist only four normed division algebras, of dimensions 1, 2, 4, and 8. And there is no 16-squares formula like Graves's eight-squares formula or Euler's four-squares formula.

Why is this? At every step along the chain of powers of two, the new number systems lose a certain amount of structure. The complex numbers are not ordered along a line. The quaternions fail to obey the algebraic rule $ab = ba$, the "commutative law." The octonions fail to obey the associative law $(ab)c = a(bc)$, though they do obey the "alternative law" $(ab)a = a(ba)$. The sedenions fail to form a division algebra and have no multiplicative norm either.

This is far more fundamental than just a failure of the Cayley–Dickson process. In 1898, Hurwitz proved that the only normed division algebras are our four old friends. In 1930, Max Zorn proved that these same four algebras are the only alternative division algebras. They truly are exceptional.

This is the sort of thing pure mathematicians, with their Platonist instincts, love. But the only really important cases for the rest of humanity seemed to be the real and complex numbers, which were of massive practical importance. The quaternions did show up in some useful if esoteric applications, but the octonions shunned the limelight of applied science.

They seemed to be a pure-mathematical dead end, the sort of pretentious intellectual nonsense you would expect from people with their heads in the clouds.

The history of mathematics shows repeatedly that it is dangerous to dismiss some clever or beautiful idea merely because it has no obvious utility. Unfortunately, this does not stop people from dismissing such ideas, often *because* they are beautiful or clever. The more "practical" people consider themselves to be, the more they tend to heap scorn on mathematical concepts that arise from abstract questions, invented "for their own sake" instead of addressing some real-world issue. The prettier the concept is, the greater the scorn, as if prettiness itself were a reason to be ashamed.

Such declarations of uselessness are hostages to fortune. It takes only one new application, one new scientific advance, and the despised concept may suddenly plonk itself down on center stage—no longer useless, but essential.

The examples are endless. Cayley himself said his matrices were completely useless, but today no branch of science could function without them. Cardano declared complex numbers to be "as subtle as they are useless," but no engineer or physicist could function in a world that lacked complex numbers. Godfrey Harold Hardy, England's leading mathematician in the 1930s, was immensely pleased that number theory had no practical application, and in particular that it could not be used in warfare. Today number theory is used to encrypt messages into code, a technique that is vital to secure Internet commerce, and even more vital to the military.

So it is turning out with the octonions. They may yet become a compulsory topic in mathematics courses, and even more so in physics. It is now emerging that the octonions are of central importance in the theory of Lie groups—especially those of interest in physics, especially the five exceptional Lie groups G_2, F_4, E_6, E_7, and E_8, with their weird dimensions 14, 52, 78, 133, and 248. Their very existence is a puzzle. One exasperated mathematician declared them a brutal act of Providence.

Lovers of nature enjoy revisiting well-known beauty spots and finding a new vantage point . . . halfway down a waterfall, just along a ledge leading off to the side of the well-worn footpath, on a promontory overlooking a blue ocean vista. In the same way, mathematicians like to revisit old topics and look at them from new points of view. As our perspective on mathematics changes, we can often reinterpret old concepts in new, insightful ways. This is not just a matter of mathematical tourism, gazing open-mouthed at the ineffable from a different angle. It provides new, powerful ways to tackle old and new problems. Nowhere has this tendency been more apparent, or more informative, than in the theory of Lie groups.

Recall that Killing organized almost all simple Lie groups into four infinite families, two of which are really parts of one larger family, the special orthogonal groups $SO(n)$ in even and odd dimensions. The other two are the special unitary groups $SU(n)$ and the symplectic groups $Sp(2n)$.

We now know that these families are all variations on the same theme. They consist of all $n \times n$ matrices satisfying a particular algebraic condition—they are "skew-Hermitian." The only difference is that you have to use matrices of real numbers to get the orthogonal Lie algebras, matrices of complex numbers to get the unitary Lie algebras, and matrices of quaternions to get the symplectic Lie algebras. These algebras come in infinite families because matrices come in infinitely many sizes. It is wonderful to see that the Lie algebras corresponding to the natural transformations of Hamilton's version of mechanics, his first great discovery, have a natural description in terms of quaternions, his last.

It does make you wonder what happens when you use octonions as entries in the matrix. Unfortunately, because of the lack of associativity, you don't get a new infinite family of simple Lie algebras. Actually, that should be "fortunately," since we know that no such family exists. But when you play the right games with octonions, and have the law of small numbers on your side, you can get Lie algebras with a vengeance.

The first hint that this might be the case came in 1914, when Élie Cartan answered an obvious question and got a surprising answer. A guiding principle in mathematics and physics is that if you have some interesting object, the first thing to ask is what its symmetry group looks like. The symmetry group of the real number system is trivial, consisting of only the identity transformation "do nothing." The symmetry group of the complex number system contains the identity and one mirror symmetry, which transforms i into $-i$. The symmetry group of the quaternions is $SU(2)$, which is very nearly the rotation group $SO(3)$ in real three-dimensional space.

Cartan asked, What is the symmetry group of the octonions?

If you are a Cartan, you can also answer this question. The symmetry group of the octonions is the smallest exceptional simple Lie group, the one known as G_2. The 8-dimensional system of octonions has a 14-dimensional symmetry group. The exceptional normed division algebra is directly related to the first of the exceptional Lie groups.

To proceed further, we need to take on board one more idea, which goes back to the Renaissance—but to the artists, not the mathematicians.

In those days, mathematics and art were rather close; not just in architecture but in painting. The Renaissance painters discovered how to apply geometry to perspective. They found geometric rules for drawing images on paper that really looked like three-dimensional objects and scenes. In so doing, they invented a new and extremely beautiful kind of geometry.

The work of earlier artists does not look quite realistic to our eyes. Even a painter like Giotto (Ambrogio Bondone) managed to produce works with an almost photographic quality, but on close analysis the perspective is not completely systematic. It was Filippo Brunelleschi who, in 1425, formulated a systematic mathematical method for obtaining accurate perspective, which he then taught to other artists. By 1435 we find the first book on the topic, Leone Alberti's *Della Pittura*.

The method was brought to perfection in the paintings of Piero della Francesca, who was also a consummate mathematician. Piero wrote three books on the mathematics of perspective. And it would be impossible not to mention Leonardo da Vinci, whose *Trattato della Pittura* begins by stating, "Let no one who is not a mathematician read my works," an echo of the slogan "Let no one ignorant of geometry enter" which legendarily was placed over the door of Plato's academy in ancient Greece.

The essence of perspective is the notion of "projection," by which a three-dimensional scene is rendered on a flat sheet of paper by (conceptually) drawing each point of the scene to the viewer's eye, and seeing where that line meets the paper. A key idea is that projections distort shapes in ways not permitted by Euclid. In particular, projection can turn parallel lines into lines that meet.

We see this effect every day. When we stand on a bridge and see a long, straight railroad track or highway disappearing into the distance, the lines converge and seem to meet at the horizon. The real lines remain the same

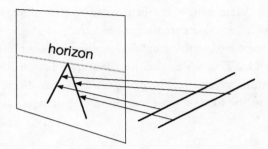

How projection makes parallels meet at the horizon.

distance apart, but perspective causes the perceived distance to shrink as the lines get farther away from us. In a mathematical idealization, infinitely long parallel lines in a plane also meet if they are suitably projected. But the place where they meet is not the image of anything in the plane—it can't be; they don't meet in the plane. It is the apparent "horizon" toward which the lines, and the plane, extend. On the plane itself, the horizon is infinitely distant, but its projection is a perfectly sensible line across the middle of the picture.

This line is known as the "line at infinity." Like the square root of minus one, it is a fiction, but an extremely useful one. The kind of geometry that emerges is called projective geometry, and in the spirit of Klein's Erlangen program, it is the geometry of those features of a scene that are not changed by projections. Every artist who makes perspective drawings with a horizon line and "vanishing points" to organize his or her images to look like real objects is using projective geometry.

In a projective plane, geometry is very elegant. Any two points can be joined by a unique line, just as in Euclid's geometry. But any two distinct lines meet, too, at exactly one point. Parallels, which so exercised Euclid, do not exist.

If this reminds you of the Fano plane, you're right. The Fano plane is a finite projective geometry.

❋

From Renaissance perspective to the exceptional Lie groups is now but a short step. The projective plane that was implicit in Alberti's methods was made explicit as a new kind of geometry. In 1636, Girard Desargues, an army officer who later became an architect and engineer, published *Pro-*

posed Draft of an Attempt to Treat the Results of a Cone Intersecting a Plane. It sounds like a book on conics, and it was, but instead of using the traditional Greek geometry, Desargues used projective methods. Just as Euclidean geometry could be turned into algebra by using Descartes's coordinates (x, y), a pair of real numbers, so projective geometry could be turned into algebra by letting x or y become infinite (in a cleverly controlled manner, involving ratios of three coordinates and setting $1 \div 0 = $ infinity).

What you can do with real numbers, you can also do with complex numbers, so now you get the complex projective plane. And if those work, why not try the quaternions or the octonions?

There are problems—the obvious methods don't work, because of the lack of commutativity. But in 1949, the mathematical physicist Pascual Jordan found a meaningful way to construct an octonionic projective plane with 16 real dimensions. In 1950, the group theorist Armand Borel proved that the second exceptional Lie group F_4 is the symmetry group of the octonionic projective plane—much like the complex plane, but formed from two 8-dimensional "rulers" labeled with octonions, not real numbers.

So now there was an octonionic explanation of two of the five exceptional Lie groups. What about the other three—E_6, E_7, and E_8?

The view of the exceptional Lie groups as brutal acts of a malicious deity was fairly widespread until 1959, when Hans Freudenthal and Jacques Tits independently invented the "magic square," and explained E_6, E_7, and E_8.

The rows and columns of the magic square correspond to the four normed division algebras. Given any two normed division algebras, you look in the corresponding row and column, and what the magic square gives you—following a technical mathematical recipe—is a Lie group. Some of these groups are straightforward; for example, the Lie group corresponding to the real row and the real column is the group SO(3) of rotations in three-dimensional space. If both row and column correspond to the quaternions, you get the group SO(12) of rotations in twelve-dimensional space, which to mathematicians is just as familiar. But if you look in the octonion row or column, the entries are the exceptional Lie groups F_4, E_6, E_7, and E_8. The missing exceptional group G_2 is also intimately associated with the octonions—as we've already seen, it is their symmetry group.

So now the general opinion is that the exceptional Lie groups exist because of the wisdom of the deity in permitting the octonions to exist. We should have known. As Einstein remarked, the Lord is subtle but not malicious. All five exceptional Lie groups are the symmetries of various octonionic geometries.

Around 1956, the Russian geometer Boris Rosenfeld, perhaps thinking about the magic square, conjectured that the three remaining exceptional groups E_6, E_7, and E_8 are also the symmetry groups of projective planes. In place of the octonions, however, you have to use the following structures:

- For E_6: the *bioctonions,* built from complex numbers and octonions.

- For E_7: the *quateroctonions,* built from quaternions and octonions.

- For E_8: the *octooctonions,* built from octonions and octonions.

The only slight snag was that no one knew how to define sensible projective planes over such combinations of number systems. But there was some evidence that the idea made sense. As matters currently stand, we can now prove Rosenfeld's conjecture, but only by making use of the groups to construct the projective planes. This is not very satisfactory, because the idea was to go the other way, from the projective planes to the groups. Still, it's a start. In fact, for E_6 and E_7 there now exist independent ways to construct the projective planes. Only E_8 is still holding out.

Were it not for the octonions, the Lie group story would be more straightforward, as Killing originally hoped, but nowhere near as interesting. Not that we mortals get to choose: the octonions, and all the associated paraphernalia, are *there.* And in some obscure way, the existence of the universe may depend on them.

The connection between the octonions and life, the universe, and everything emerges from string theory. The key feature is the need for extra dimensions to hold the strings. Those extra dimensions can in principle have lots of shapes, and the big question is to find the *right* shape. In old-fashioned quantum theory, a key principle is symmetry, and that's the case in string theory too. So of course Lie groups get in on the act. Everything hinges on those Lie groups of symmetries, and again the exceptional groups stick out—not as sore thumbs, but as opportunities for unusual coincidences that could help make the physics work.

Which gets us back to the octonions.

Here's an example of their influence. In the 1980s, physicists noticed that a rather nice relationship occurs in space-times of 3, 4, 6, and 10 dimensions. Vectors (directed lengths) and spinors (algebraic gadgets originally created by Paul Dirac in his theory of electron spin) are very neatly related in these dimensions, and only these. Why? It turns out that the vector–spinor relationship holds precisely when the dimension of space-time is 2 greater than that of a normed division algebra. Subtract 2 from 3, 4, 6, and 10, and what you get is 1, 2, 4, and 8.

The mathematical point is that in 3-, 4-, 6-, and 10-dimensional string theory, every spinor can be represented using two numbers in the associated normed division algebra. This doesn't happen for any other number of dimensions, and it has a number of nice consequences for physics. So we have four candidate string theories here: real, complex, quaternionic, and octonionic. And it so happens that among these possible string theories, the one that is currently thought to have the best chance of corresponding to reality is the 10-dimensional one, *specified by the octonions*. If this 10-dimensional theory really does correspond to reality, then our universe is built from octonions.

And that's not the only place where these strange "numbers," barely clinging to that name because they just satisfy enough of the rules of algebra, are influential. That fashionable new candidate string theory, M-theory, involves 11-dimensional space-time. In order to reduce the perceptible part of space-time from 11 dimensions to the familiar 4, we have to throw away 7 by rolling them up so tightly that they can't be detected. And how do you do that for 11-dimensional supergravity? You make use of the exceptional Lie group G_2, the symmetry group of the octonions.

There they are again: no longer quaint Victoriana but a hefty clue to a possible Theory of Everything. It's an octonionic world.

16

SEEKERS AFTER TRUTH AND BEAUTY

Was Keats right? Is beauty truth, and truth beauty?

The two are intimately connected, possibly because our minds react similarly to both. But what works in mathematics need not work in physics, and vice versa. The relationship between mathematics and physics is deep, subtle, and puzzling. It is a philosophical conundrum of the highest order—how science has uncovered apparent "laws" in nature, and why nature seems to speak in the language of mathematics.

Is the universe genuinely mathematical? Are its apparent mathematical features mere human inventions? Or does it seem mathematical to us because mathematics is the deepest aspect of its infinitely complex nature that we are able to understand?

Mathematics is not some disembodied version of ultimate truth, as many used to think. If anything emerges from our tale, it is that mathematics is created by people. We can readily identify with their triumphs and their tribulations. Who could fail to be moved by the appalling deaths of Abel and Galois, both at the age of 21? One was deeply loved but never earned enough money to marry; the other, brilliant and unstable, fell in love but was rejected, and perhaps died because of that love. Today's medical advances would have saved Abel, and might even have helped Hamilton stay sober.

Because mathematicians are human and live ordinary human lives, the creation of new mathematics is partly a social process. But neither mathematics nor science is *wholly* the result of social processes, as social relativists often claim. Both must respect external constraints: logic, in the case of mathematics, and experiment, in the case of science. However desperately mathematicians might want to trisect an angle by Euclidean methods, the plain fact is that it is impossible. However strongly physicists might want Newton's law of gravity to be the ultimate description of the universe, the motion of the perihelion of Mercury proves that it's not.

This is why mathematicians are so stubbornly logical, and obsessed by concerns that most people could not care less about. Does it really *matter* whether you can solve a quintic by radicals?

History's verdict on this question is unequivocal. It does matter. It may not matter directly for everyday life, but it surely matters to humanity as a whole—not because anything important rests on being able to solve quintic equations, but because understanding why we can't opens a secret doorway to a new mathematical world. If Galois and his predecessors had not been obsessed with understanding the conditions under which an equation can be solved by radicals, humanity's discovery of group theory would have been greatly delayed, and perhaps might not have happened.

You may not encounter groups in your kitchen or on your drive to work, but without them today's science would be severely curtailed, and our lives would be far different. Not so much in gadgetry like jumbo jets or GPS navigation or cell phones—though those are part of the story too—but in insight into nature. No one could have predicted that a pedantic question about equations could reveal the deep structure of the physical world, but that is what happened.

The clear message of history is a simple one. Research on deep mathematical issues should not be rejected or denigrated merely because those issues seem to have no direct practical use. Good mathematics is more valuable than gold, and where it comes from is mostly irrelevant. What counts is where it leads.

The astonishing thing is that the best mathematics usually leads somewhere unexpected, and a lot of it turns out to be vital for science and technology, even though it was originally invented for some totally different purpose. The ellipse, studied by the Greeks as a section of a cone, was the clue that led, via Kepler, from Tycho Brahe's observations of the motion of Mars to Newton's theory of gravity. Matrix theory, whose inventor Cayley apologized for its uselessness, became an essential tool in statistics, economics, and virtually every branch of science. The octonions may be the inspiration for a Theory of Everything. Of course, the theory of superstrings may turn out to be just a pretty piece of mathematics with no relevance to physics. If so, the existing uses of symmetry in quantum theory still demonstrate that group theory provides deep insights into nature, even though it was developed to answer a question in pure mathematics.

Why is mathematics so useful for purposes that its inventors never intended?

The Greek philosopher Plato said that "God ever geometrizes." Galileo said much the same thing: "Nature's great book is written in mathematical language." Johannes Kepler set out to find mathematical patterns in planetary orbits. Some of them led Newton to his law of gravitation; others were mystical nonsense.

Many modern physicists have commented on the astonishing power of mathematical thought. Wigner alluded to the "unreasonable effectiveness" of mathematics as a way to understand nature; the phrase appears in the title of an article he wrote in 1960. In the body of the article he said he would tackle two main points:

> The first point is that the enormous usefulness of mathematics in the natural sciences is something bordering on the mysterious and that there is no rational explanation for it. Second, it is just this uncanny usefulness of mathematical concepts that raises the question of the uniqueness of our physical theories.

And:

> The miracle of the appropriateness of the language of mathematics for the formulation of the laws of physics is a wonderful gift which we neither understand nor deserve. We should be grateful for it and hope that it will remain valid in future research and that it will extend, for better or for worse, to our pleasure, even though perhaps also to our bafflement, to wide branches of learning.

Paul Dirac believed that in addition to being mathematical, nature's laws also had to be beautiful. In his mind, beauty and truth were two sides of the same coin, and mathematical beauty gave a strong clue to physical truth. He even went so far as to say he would prefer a beautiful theory to a correct one, and that he valued beauty above simplicity: "The research worker, in his efforts to express the fundamental laws of nature in mathematical form, should strive mainly for mathematical beauty. He should still take simplicity into consideration in a subordinate way to beauty . . . where they clash the latter must take precedence."

Interestingly, Dirac's concept of beauty in mathematics differed considerably from that of most mathematicians. It did not include logical rigor, and many steps in his work had logical gaps—the best-known example being his "delta function," which has self-contradictory properties. Nevertheless, he made very effective use of this "function," and eventually mathematicians reformulated the idea rigorously—at which point it was indeed a thing of beauty.

Still, as Dirac's biographer Helge Kragh has remarked, "All of [Dirac's] great discoveries were made before [the mid-1930s], and after 1935 he largely failed to produce physics of lasting value. It is not irrelevant to point out that the principle of mathematical beauty governed his thinking only during the later period."

Not irrelevant, perhaps, but not correct either. Dirac may have made the principle explicit during his later period, but he was using it earlier. *All* of his best work is mathematically elegant, and he relied on elegance as a test of whether he was heading in a fruitful direction. What all this suggests is not that mathematical beauty is *the same as* physical truth but that it is *necessary* for physical truth. It is not sufficient. Many beautiful theories have turned out, once confronted with experiments, to be complete nonsense. As Thomas Huxley said, "Science is organized common sense, where many a beautiful theory was killed by an ugly fact."

Yet there is much evidence that nature, at root, is beautiful. The mathematician Hermann Weyl, whose research linked group theory and physics, said, "My work has always tried to unite the true with the beautiful and when I had to choose one or the other, I usually chose the beautiful." Werner Heisenberg, a founder of quantum mechanics, wrote to Einstein,

You may object that by speaking of simplicity and beauty I am introducing aesthetic criteria of truth, and I frankly admit that I am strongly attracted by the simplicity and beauty of the mathematical schemes which nature presents us. You must have felt this too: the almost frightening simplicity and wholeness of the relationship, which nature suddenly spreads out before us.

Einstein, in turn, felt that so many fundamental things are unknown—the nature of time, the sources of ordered behavior of matter, the shape of the universe—that we must remind ourselves how far we are from understanding anything "ultimate." To the extent that it is useful, mathematical elegance provides us only local and temporary truths. Still, it is our best way forward.

Throughout history, mathematics has been enriched from two different sources. One is the natural world, the other the abstract world of logical thought. It is these two in combination that give mathematics its power to inform us about the universe. Dirac understood this relationship perfectly: "The mathematician plays a game in which he himself invents the rules while the physicist plays a game in which the rules are provided by nature, but as time goes on it becomes increasingly evident that the rules which the mathematician finds interesting are the same as those which nature has chosen." Pure and applied mathematics complement each other. They are not poles apart, but the two ends of a connected spectrum of thought.

The story of symmetry demonstrates how even a negative answer to a good question ("can we solve the quintic?") can lead to deep and fundamental mathematics. What counts is *why* the answer is negative. The methods that reveal this can be used to solve many other problems—among them, deep questions in physics. But our story also shows that the health of mathematics depends on the infusion of new life from the physical world.

The true strength of mathematics lies precisely in this remarkable fusion of the human sense of pattern ("beauty") with the physical world, which acts both as a reality check ("truth") and as an inexhaustible source of inspiration. We cannot solve the problems posed by science without new mathematical ideas. But new ideas for their own sake, if carried to extremes, can degenerate into meaningless games. The demands of science keep mathematics running along fruitful lines, and frequently suggest new ones.

If mathematics were entirely demand-driven, a slave of science, you would get the work you expect from a slave—sullen, grudging, and slow. If the subject were entirely driven by internal concerns, you would get a spoilt, selfish brat—pampered, self-centered, and full of its own importance. The best mathematics balances its own needs against those of the outside world.

This is what its unreasonable effectiveness derives from. A balanced personality learns from its experiences, and transfers that learning to new circumstances. The real world inspired great mathematics, but great mathematics can transcend its origins.

The unknown Babylonian who discovered how to solve a quadratic equation could never have realized, in his wildest dreams, what his legacy would be more than three thousand years later. No one could have predicted that questions about the solvability of equations would lead to one of the core concepts of mathematics, that of a group, or that groups would prove to be the language of symmetry. Even less could anyone have known that symmetry would unlock the secrets of the physical world.

Being able to solve a quadratic has very limited utility in physics. Being able to solve a quintic is even less useful, if only because any solution must be numerical, not symbolic, or else employ symbols specially invented for the purpose, which do little more than cover the question with a fig leaf. But understanding why quintics cannot be solved, appreciating the crucial role of symmetry, and pushing the underlying idea as far as it could go—that opened up entire new physical realms.

The process continues. The implications of symmetry for physics, indeed for the whole of science, are still relatively unexplored. There is much that we do not yet understand. But we do understand that symmetry groups are our path through the wilderness—at least until a still more powerful concept (perhaps already waiting in some obscure thesis) comes along.

In physics, beauty does not automatically ensure truth, but it helps.

In mathematics, beauty *must* be true—because anything false is ugly.

FURTHER READING

John C. Baez, "The octonions," *Bulletin of the American Mathematical Society* volume 39 (2002) 145–205.

E. T. Bell, *Men of Mathematics* (2 volumes), Pelican, Harmondsworth, 1953.

R. Bourgne and J.-P. Azra, *Écrits et Mémoires Mathématiques d'Évariste Galois,* Gauthier-Villars, Paris, 1962.

Carl B. Boyer, *A History of Mathematics,* Wiley, New York, 1968.

W. K. Bühler, *Gauss: A Biographical Study,* Springer, Berlin, 1981.

Jerome Cardan, *The Book of My Life* (translated by Jean Stoner), Dent, London, 1931.

Girolamo Cardano, *The Great Art or the Rules of Algebra* (translated T. Richard Witmer), MIT Press, Cambridge, MA, 1968.

A. J. Coleman, "The greatest mathematical paper of all time," *The Mathematical Intelligencer,* volume 11 (1989) 29–38.

Julian Lowell Coolidge, *The Mathematics of Great Amateurs,* Dover, New York, 1963.

P. C. W. Davies and J. Brown, *Superstrings,* Cambridge University Press, Cambridge, 1988.

Underwood Dudley, *A Budget of Trisections,* Springer, New York, 1987.

Alexandre Dumas, *Mes Mémoires* (volume 4), Gallimard, Paris, 1967.

Euclid, *The Thirteen Books of Euclid's Elements* (translated by Sir Thomas L. Heath), Dover, New York, 1956 (3 volumes).

Carl Friedrich Gauss, *Disquisitiones Arithmeticae* (translated by Arthur A. Clarke), Yale University Press, New Haven, 1966.

Jan Gullberg, *Mathematics: From the Birth of Numbers,* Norton, New York, 1997.

George Gheverghese Joseph, *The Crest of the Peacock,* Penguin, London, 2000.

Brian Greene, *The Elegant Universe,* Norton, New York, 1999.

Michio Kaku, *Hyperspace,* Oxford University Press, Oxford, 1994.

Morris Kline, *Mathematical Thought from Ancient to Modern Times,* Oxford University Press, Oxford, 1972.

Helge S. Kragh, *Dirac—A Scientific Biography,* Cambridge University Press, Cambridge, 1990.

Mario Livio, *The Equation That Couldn't Be Solved,* Simon & Schuster, New York, 2005.

J.-P. Luminet, *Black Holes,* Cambridge University Press, Cambridge, 1992.

Oystein Ore, *Niels Henrik Abel: Mathematician Extraordinary,* University of Minnesota Press, Minneapolis, 1957.

Abraham Pais, *Subtle Is the Lord: The Science and the Life of Albert Einstein,* Oxford University Press, Oxford, 1982.

Roger Penrose, *The Road to Reality,* BCA, London, 2004.

Lisa Randall, *Warped Passages,* Allen Lane, London, 2005.

Michael I. Rosen, "Niels Hendrik Abel and equations of the fifth degree," *American Mathematical Monthly* volume 102 (1995) 495–505.

Tony Rothman, "The short life of Évariste Galois," *Scientific American* (April 1982) 112–120. Collected in Tony Rothman, *A Physicist on Madison Avenue,* Princeton University Press, 1991.

H. F. W. Saggs, *Everyday Life in Babylonia and Assyria,* Putnam, New York, 1965.

Lee Smolin, *Three Roads to Quantum Gravity,* Basic Books, New York, 2000.

Paul J. Steinhardt and Neil Turok, "Why the cosmological constant is small and positive," *Science* volume 312 (2006) 1180–1183.

Ian Stewart, *Galois Theory* (3rd edition), Chapman and Hall/CRC Press, Boca Raton 2004.

Jean-Pierre Tignol, *Galois's Theory of Algebraic Equations,* Longman, London, 1980.

Edward Witten, "Magic, mystery, and matrix," *Notices of the American Mathematical Society* volume 45 (1998) 1124–1129.

WEBSITES

A. Hulpke, Determining the Galois group of a rational polynomial: http://www.math.colosate.edu/hulpke/talks/galoistalk.pdf

The MacTutor History of Mathematics archive: http://www-history.mcs.st-andrews.ac.uk/index.html

A. Rothman, Genius and biographers: the fictionalization of Évariste Galois: http://godel.ph.utexas.edu/tonyr/galois.htm

INDEX